Representation Discovery using Harmonic Analysis

Synthesis Lectures on Artificial Intelligence and Machine Learning

Editors
Ronald J. Brachman, *Yahoo! Research*

Thomas Dietterich, *Oregon State University*

Representation Discovery using Harmonic Analysis
Sridhar Mahadevan
2008

Essentials of Game Theory: A Concise, Multidisciplinary Introduction
Kevin Leyton-Brown and Yoav Shoham
2008

A Concise Introduction to Multiagent Systems and Distributed Artificial Intelligence
Nikos Vlassis
2007

Intelligent Autonomous Robotics
Peter Stone
2007

Representation Discovery using Harmonic Analysis
Sridhar Mahadevan

ISBN: 978-3-031-00418-6   paperback
ISBN: 978-3-031-01546-5   ebook

DOI 10.1007/978-3-031-01546-5

A Publication in the Springer series

SYNTHESIS LECTURES ON ARTIFICIAL INTELLIGENCE AND MACHINE LEARNING #4

Lecture #4
Series Editor: Ronald J. Brachman, Yahoo! Research and Thomas Dietterich, Oregon State University

Series ISSN: 1939-4608   print
Series ISSN: 1939-4616   electronic

First Edition
10 9 8 7 6 5 4 3 2 1

REPRESENTATION DISCOVERY USING HARMONIC ANALYSIS

Sridhar Mahadevan

Computer Science Department,

University of Massachusetts, Amherst, MA, USA

SYNTHESIS LECTURES ON ARTIFICIAL INTELLIGENCE AND MACHINE LEARNING #4

ABSTRACT

Representations are at the heart of artificial intelligence (AI). This book is devoted to the problem of representation discovery: how can an intelligent system construct representations from its experience? Representation discovery re-parameterizes the state space – prior to the application of information retrieval, machine learning, or optimization techniques – facilitating later inference processes by constructing new task-specific bases adapted to the state space geometry. This book presents a general approach to representation discovery using the framework of harmonic analysis, in particular Fourier and wavelet analysis. Biometric compression methods, the compact disc, the computerized axial tomography (CAT) scanner in medicine, JPEG compression, and spectral analysis of time-series data are among the many applications of classical Fourier and wavelet analysis. A central goal of this book is to show that these analytical tools can be generalized from their usual setting in (infinite-dimensional) Euclidean spaces to discrete (finite-dimensional) spaces typically studied in many subfields of AI. Generalizing harmonic analysis to discrete spaces poses many challenges: a discrete representation of the space must be adaptively acquired; basis functions are not pre-defined, but rather must be constructed. Algorithms for efficiently computing and representing bases require dealing with the curse of dimensionality. However, the benefits can outweigh the costs, since the extracted basis functions outperform parametric bases as they often reflect the irregular shape of a particular state space. Case studies from computer graphics, information retrieval, machine learning, and state space planning are used to illustrate the benefits of the proposed framework, and the challenges that remain to be addressed. Representation discovery is an actively developing field, and the author hopes this book will encourage other researchers into exploring this exciting area of research.

KEYWORDS

Artificial intelligence, Dimensionality reduction, Feature construction, Harmonic analysis, Image processing, Information retrieval, Linear algebra, Machine learning, Natural language processing, State space planning

Contents

Preface

Many successes of artificial intelligence (AI) have relied on human expertise to hand-craft a set of task-specific *features* that map the original state space – collections of images, problem states, or words – into an implicit vector space. This book investigates the problem of automating *representation discovery*: the development of a computational framework for constructing *features* or *basis functions* from data well-suited to solving a particular task or range of tasks – such as compression, information retrieval, learning, and planning – on a given state space. This book presents a mathematically principled approach to representation discovery based on the framework of *harmonic analysis*. The fundamental idea in harmonic analysis is to map phenomena that occur over space and time into a *frequency*-oriented coordinate system.

Harmonic analysis dates back to 1807 when Joseph Fourier, in the course of solving the heat equation, made a remarkable discovery that arbitrary real-valued functions could be decomposed as linear combinations of highly symmetric trigonometric functions. Fourier analysis has since played a central role in mathematics, science and technology. It connects continuous mathematics, such as linear differential equations, to concepts in discrete mathematics, such as linear algebra and matrix theory, using the principle of *diagonalization*. It has lead to many applications, from the compact disc and JPEG image compression to the computerized axial tomography (CAT) scanner in medicine. Despite these successes, two centuries of research into Fourier analysis has revealed significant chinks in its armor: the basis functions are global, making it difficult to represent piecewise-smooth functions with local discontinuities. Fourier analysis also does not easily reveal multiscale regularities. Addressing these challenges has taken the combined efforts of engineers, mathematicians, and scientists, giving rise to a new mathematical microscope for probing the properties of functions. This new paradigm is based on the theory of *wavelets*. Diagonalization is replaced by the concept of *dilation*, where basis elements are constructed at multiple levels of spatial and temporal abstraction. Unlike Fourier analysis where time or space is mapped into frequencies, wavelet analysis adaptively combines time and space into time-frequency or space-frequency atoms of varying granularities.

Harmonic analysis provides a powerful framework for representation discovery in many areas of AI, from classification and compression to optimization and information retrieval. A key strength of harmonic analysis is that it produces an ordered and meaningful summarization of the underlying data or state space. Harmonic analysis extracts regularities from data by projecting them into *invariant subspaces*. Many problems in AI are naturally defined on discrete state spaces, such as graphs. Here, Fourier and wavelet bases are not pre-defined, but rather have to be discovered from samples of the underlying data or state space. Fourier analysis in discrete spaces is based on diagonalizing a discrete version of the "Laplacian" operator from continuous spaces. This operator, commonly termed the graph Laplacian, has recently emerged as a key object of interest in a number of areas, including dimensionality reduction, Markov processes, spectral graph theory, and web page ranking.

While Fourier analysis is starting to play a more prominent role in discrete spaces, wavelets have rarely been used in "mainstream" AI (a notable exception being David Marr's pioneering work in computer vision). A major goal of this book is to introduce a novel and highly promising wavelet framework for multiscale analysis, where basis functions emerge from the dilatory actions of a diffusion operator on the graph, such as the random walk. A major challenge of using harmonic analysis is its computational complexity in large discrete and continuous spaces. Computing eigenvectors or wavelets can be intractable: the book describes several promising approaches to scaling harmonic analysis by combining matrix compression, sampling, and domain knowledge. A number of case studies, from computer graphics to natural language processing and state space planning, are used to illustrate the core concepts and the range of possible applications.

This book summarizes research done in collaboration with many researchers. Professor Mauro Maggioni of the Mathematics and Computer Science Departments at Duke University has been a close collaborator, and I am indebted to him for his generous assistance. My PhD students – Kimberly Ferguson, Mohammad Ghavamzadeh, Jeff Johns, Vimal Mathew, Sarah Osentoski, Khashayar Rohanimanesh, and Chang Wang – as well as other members of the PVF group, notably Marek Petrik and Illya Scheidwasser, contributed to the research in innumerable ways. The Autonomous Learning Laboratory at the University of Massachusetts, Amherst, provided a stimulating environment. I thank its current and former PhD students, and its co-director Professor Andrew Barto, for helpful feedback. Finally, support for the research described in this book was provided in part by the National Science Foundation under grant IIS-0534999.

Sridhar Mahadevan
Amherst, Massachusetts

CHAPTER 1

Overview

Representations have long played a leading role in artificial intelligence (AI). Much effort has gone into devising representations in specific subfields, from perception and problem-solving to decision-making and robotics. The goal of designing agents that can discover novel representations from their environment has been a longstanding challenge. Amarel [2] pioneered the view that agents should analyze state spaces to determine geometric properties such as "bottlenecks" and "symmetries". This book presents a unified framework for *representation discovery*—the construction of a set of *basis functions* that capture the regularities of a particular state space, and facilitate tasks such as compression, learning, and planning. To limit its scope, the notion of representation in this book adheres closely to its usage in mathematics: a representation of a set, such as a vector space, is a (usually small) number of *orthogonal elements*—or a *basis*—that can be used to generate in a unique manner all the remaining elements. This simple notion of representation turns out to be surprisingly rich both theoretically—it includes representations in finite-dimensional linear algebra [110], group theory [51], and infinite-dimensional function spaces [37]—as well as in terms of applications, ranging from machine learning [16, 104], Markov decision processes [98] and reinforcement learning [111], perception [106], computer graphics [59], and information retrieval [36].

This book presents a general approach to automatic basis construction using the framework of *harmonic analysis* [5, 25, 50]. Harmonic analysis is a field of mathematics that is two centuries old: it dates at least back to 1807 when Joseph Fourier conjectured that arbitrary real-valued functions could be decomposed into elementary combinations of highly regular trigonometric functions. Fourier analysis has played a central role in mathematics, science, and technology ever since, dominating many intellectual developments in these areas as well as many successful applications. The compact disc, JPEG compression [3, 122], and the computerized axial tomography (CAT) scanner in medicine are but a few of the hundreds of commercial applications of Fourier analysis. For all its successes, Fourier analysis has some significant limitations: it is difficult to represent certain classes of functions, and the analysis did not reveal multiscale regularities. To address these challenges, a new paradigm emerged over the past two decades based on the theory of *wavelets* [34, 81, 82]. Here, the basis elements are constructed at multiple levels of abstraction, and unlike Fourier analysis where time or space is mapped into frequencies, wavelet analysis is based on combining time and space. Instead of differential equations, which give rise to Fourier analysis, wavelets are based on dilation equations.

Much work in machine learning [16] is based on *likelihood analysis*, where a parametric generative model $P_\theta(X)$ is used to fit the data X by finding the most likely setting of the parameters θ using the likelihood function $L(\theta)$. In contrast, harmonic analysis is based on finding a projection of the data X onto a set of basis functions Φ, which span a set of *invariant* subspaces. A simple example of an invariant subspace in vector spaces is the one-dimensional space spanned by an eigenvector associated with a specific eigenvalue of a matrix T. For example, one form of Fourier analysis on graphs is to diagonalize the Laplacian matrix $L = D - A$, where A is the adjacency matrix and D is the *valency* matrix of row sums of A. The strength of harmonic analysis is that it produces an ordered and meaningful summarization of the underlying data or state space, since typically the basis functions can be ordered in terms of *smoothness*. The first eigenvector of the Laplacian matrix L is the constant function, and projecting the data X onto the first eigenvector produces the *average* or mean value. Thus, low-order basis functions capture the "low-frequency" components of the data, and higher-order basis functions progressively fill in details.

A major challenge of using harmonic analysis is its computational complexity in large discrete and continuous spaces. Computing eigenvectors can be intractable: the book describes several promising approaches that enable harmonic analysis to be scaled to large discrete spaces using matrix compression, sampling, and domain knowledge.

1.1   WHAT IS A REPRESENTATION?

At the outset, it is important to precisely clarify what we mean by a *representation*. It is a formalism for describing a class of objects with respect to a well-defined *basis* [110]. The simplest example of representation is that of a number system: 3, III, and 011 all refer to the same object, namely the number three. These alternate representations arise—decimal, Roman, and binary—from a different choice of a *basis*. The choice of a basis can have a dramatic impact on the efficiency of computation, as well as in the amount of storage taken up by a representation, and above all, in revealing deeper properties of the set as a whole.

The selection of a basis is of crucial importance: the decimal number system is a notable example. In a nutshell, *this book is about the theory and algorithms for automatic basis construction*. This endeavor is in many ways a new field of research within AI: representation discovery unifies issues that would otherwise be somewhat disparate. As we will see, the same issues of basis construction arise regardless of whether the final objective is machine learning, optimization, or search. Also, the objects being represented can be diverse: they range from states and actions in discrete or continuous state spaces in tasks from robotics to games, or training data such as documents or images for an information-retrieval or multimedia system.

The *place-value* notation, a hallmark of decimal representations, was a landmark in human representation discovery. The two occurrences of 3 in the decimal representation 323 have a very different meaning: the first occurrence of 3—reading from the right—means the number "three". The second occurrence means a number one hundred times larger—meaning, the number "three hundred". The number 323 has a *linear expansion* in terms of the underlying *basis functions* (here, powers of 10):

$$323 = 3(10)^2 + 2(10)^1 + 3(10)^0.$$

Each coefficient in this expansion can be determined by dividing the number by the relevant basis function:

$$323 \div (10^2) = 3, \quad 23 \div (10) = 2, \quad 3 \div (10^0) = 3.$$

Thus, a number can be *analyzed* into its respective building blocks, as well as *synthesized* by recombining these building blocks in the right manner. This *analysis–synthesis* perspective forms a unifying theme for our book, and will be extensively used in subsequent chapters.

1.2   PRINCIPLES OF REPRESENTATION DISCOVERY

Representation discovery is facilitated by a change in basis: the example of human discovery of decimal numbers was enabled by the development of the base 10 representation. Are there broad principles underlying basis change that can be exploited in terms of designing suitable algorithms for basis construction? In this book, we explore two broad principles, both stemming from the field of harmonic analysis [50, 81]:

- *Space or time → frequency:* One generic principle for basis construction involves remapping functions over time or space into a frequency-oriented coordinate system, generically termed *Fourier* analysis. Examples include dimensionality reduction methods such as principal components analysis (PCA) [57] and singular value decomposition (SVD) [47], time-series and image-compression methods such as Fourier transforms [81] and JPEG [122], and recent *manifold* and *graph*-based methods, such as diffusion maps [28], ISOMAP [114], LLE [101], and Laplacian eigenmaps [90].

- *Space or time → multiscale space-frequency or time-frequency atoms:* A more powerful basis conversion process involves a multiscale construction where functions over space or time are progressively remapped into time-frequency or space-frequency *atoms* [34, 81]. This multiscale construction is most characteristic of a family of more recent methods called *wavelets* [34, 81]. We will explore multiscale basis construction on graphs and manifolds using a recent graph-based approach called *diffusion wavelets* [30].

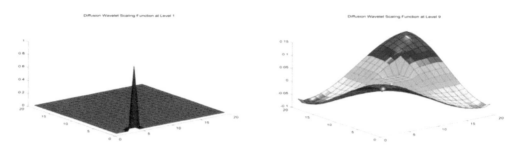

FIGURE 1.1: Harmonic analysis involves a basis conversion remapping functions over time or space into a frequency-oriented representation. There are intrinsic theoretical limitations that prevent information from being localized in both time/space and frequency. Left: a unit delta function, which is localized in space, but covers all frequencies. Right: an eigenvector, which is localized in frequency, but its support is global across the space.

There are intrinsic theoretical limitations that govern basis change: a generalization of the *Heisenberg uncertainty principle* states that the distribution of "energy" of a function (over time or space) and its Fourier transform (over frequency) cannot be simultaneously arbitrarily small [81]. Figure 1.1 illustrates this paradox. The figure shows a graph representing a spatial environment with a centrally located "obstacle" region representing inaccessible vertices. A unit delta function is localized in space (the unit vector is 1 on exactly one vertex and 0 everywhere else), but the space of such functions covers all frequency bands. In contrast, an eigenvector is localized in frequency, but spatially its support is global. Managing this tradeoff between information localization in space/time and frequency is one of the central challenges governing basis conversion.

1.3 OVERVIEW OF THE BOOK

Figure 1.2 illustrates a framework for representation discovery that broadly summarizes the approach described in this book. Representation discovery (alternatively "feature discovery" or "basis construction") is a process that lies in between data collection or state space exploration in a given domain, and tasks such as the analysis of data [16] or optimization [98]. That is, basis selection *precedes* the application of standard data analysis or optimization methods studied in conventional machine learning [16] or AI [102], such as clustering [88], classification [90], regression [91], and stochastic state space planning [98]. Even more significantly, the construction of a basis can be done *independently* of the specific problem being solved.

Of course, it is possible to modify the basis construction process so that it is sensitive to the final task or the function being approximated, and as we will see, such task-specific basis construction can yield better results. However, much of our presentation of the theory of

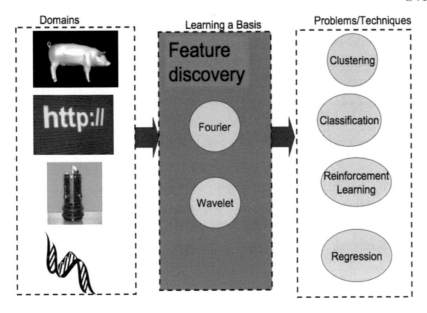

FIGURE 1.2: A framework for representation discovery

basis construction will ignore specifics related to the particular problem or method. The basis construction process may depend on the structure of the initial samples, since for example, one step involves the construction of a *graph* from the data samples. This graph depends on a *local neighborhood* relationship, which can be domain-specific.

As Figure 1.2 illustrates, the two major types of basis construction approaches described in this book are *Fourier* and *wavelet* bases. These two types of bases arise in a field of mathematics called *harmonic analysis* [5, 25, 50]. Its history dates back to 1807, when Joseph Fourier discovered a general representation for arbitrary real-valued functions using highly symmetric basis functions—the trigonometric functions. Fourier's insight was to transform data defined over time or space into a frequency-oriented representation. This insight has had a lasting impact in mathematics, science, and technology for two centuries. Fourier analysis has been the mainstay not only in scientific fields ranging from particle physics and molecular chemistry, but also played a leading role in engineering and technology from signal processing and consumer electronics, to time-series analysis (e.g., weather prediction) and medical devices such as CAT scanners.

Nevertheless, gradually over the past century, limitations of Fourier analysis were discovered, and through the efforts of a wide group of researchers in mathematics and engineering, an alternate representation emerged based on the theory of *wavelets* [34, 81]. Here, basis functions are not associated with frequencies, but rather with a multiscale decomposition of time (or space) and frequency taken together. Fourier and wavelet analysis constitute the most important building blocks of harmonic analysis. This field is closely associated with the study of *symmetry*, or group theory [51]. We briefly review the main concepts in these areas, showing how they generalize to discrete spaces of interest in AI. This generalization provides the foundation for a new perspective on representation discovery. Generalizing Fourier and wavelet analysis from Euclidean spaces to non-Euclidean spaces defined by graphs, groups, and manifolds enables new basis discovery techniques to be developed in discrete data and search spaces.

1.3.1 Road Map to the Book

We give a brief road map to the book, describing each of the major parts in more detail below. Broadly speaking, the book is divided into three parts: theoretical foundations, algorithms and computational tractability, and applications and case studies.

1.3.2 Theory of Basis Construction: Vector Spaces

In the example above on human discovery of decimal representations, we introduced the analysis–synthesis perspective, which will inform much of the book. In Chapter 2, we introduce some of the theoretical foundations of representation discovery, building on matrix representation theory [110]. Here, we will see how the selection of a basis is crucial for an efficient representation of linear mappings on vector spaces. A fundamental concept that we will use extensively is the notion of an *invariant subspace*: these are spaces spanned by vectors that remain invariant under the application of a linear operator (e.g., a matrix). An example of an invariant subspace is an *eigenspace*, which is the space of vectors associated with an eigenvalue. Basis construction can be formalized as the process of finding invariant subspaces. We will also introduce an abstraction of the analysis–synthesis perspective, which is an abstraction of the process of basis construction. In particular, an object x (e.g., a function on a graph) is synthesized in terms of a set of basis functions $B = \{\phi_1, \phi_2, \ldots\}$ by the linear expansion:

$$x = \sum_i \alpha_i \phi_i,$$

where each coefficient α_i can be viewed as a "measurement" of the object. These measurements will be abstractly denoted by *linear functionals* of the form:

$$\alpha_i = \langle x, \psi_i \rangle,$$

where $\langle x, \psi_i \rangle$ represents an *inner product* in a *vector space* [37], and ψ is a *dual basis* to ϕ. As we will show in this book, this mathematical formulation is surprisingly rich, and covers many interesting and real-world applications of AI (described below in Section 1.3.5).

1.3.3 Generalizing Fourier and Wavelet Analysis

Classical Fourier and wavelet analysis is usually in the context of Euclidean spaces, e.g. the space of n-dimensional real vectors or $\mathbb{R}^n$ [34, 82]. An important goal of this book is to show how much of the fundamental theory underlying these approaches generalizes to discrete spaces like graphs of central interest to AI and computer science. In Chapter 3, we describe Fourier analysis on graphs, where basis functions span invariant eigenspaces of random walk operators on graphs. The graph Laplacian [26] plays a central role in the construction of basis functions on graphs, due to its close connection both to classical Fourier analysis and the continuous Laplacian on manifolds [99], as well as its relationship to random walks. The eigenvectors of the graph Laplacian reveal a surprising amount of information about a graph: the second *Fiedler* eigenvector [44] is often used to partition a graph.

In Chapter 4, we introduce multi-resolution analysis on graphs, generalizing the theory of wavelets [34, 82] developed in Euclidean spaces to the discrete space of graphs. Wavelets address many of the limitations of Fourier analysis, in particular the basis functions are localized *both* in time and frequency, and the analysis is intrinsically at multiple spatial and temporal resolutions. The core new idea in wavelets is that of *dilation*: basis functions are constructed out of simpler bases by stretching them in time or space. On a graph, this takes on an interesting new interpretation. Chapter 4 introduces the idea of *diffusion wavelets* [30], where the basis functions are constructed at each level by dilating the ones at the previous level using powers of the random walk on a graph. This process constructs a hierarchy of vector spaces, and two sets of basis functions called *scaling functions* representing coarser views, and *wavelets*, representing the finer detailed view.

1.3.4 Algorithms and Computational Tractability

Computational tractability is one of the crucial concerns that needs to be addressed in applying harmonic analysis to representation discovery. How expensive is the process of constructing basis functions? Chapter 5 explores some ideas for scaling basis construction to large continuous and discrete spaces. We will also introduce the idea of Kronecker product and sum representations, such as the Kronecker product of two graphs [31] or matrices [120]. Chapter 5 generalizes the Laplacian from graphs to manifolds [99], subsets of Euclidean space that can only locally be viewed as Euclidean. Manifolds are of growing importance in AI, in particular in machine learning where the space of parametric probability distributions forms a manifold [66]. The Laplacian on a manifold is introduced [99], and a general result called Hodge theorem is stated that shows the eigenfunctions of the Laplacian on a manifold provide a complete discrete basis for all square integrable functions on a continuous manifold. The key problem in extending eigenfunctions to continuous spaces is how to extend sample values to new unobserved values:

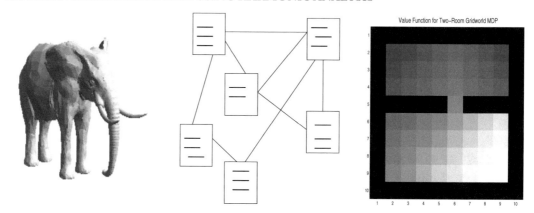

FIGURE 1.3: Three challenging application problems for evaluating the representation discovery algorithms described in this book. Left: a 3D object in computer graphics. Middle: a collection of text documents; Right: a stochastic state-space planning problem.

this is the problem of *out-of-sample* extension. One solution based on *Nyström* interpolation is described [41].

1.3.5   Case Studies

Figures 1.3 and 1.4 illustrate some of the major application areas that will be studied in this book, as well as example basis functions that were "discovered" using the algorithms described in this book. These examples come from a series of case studies, which form the third and final phase of this book, illustrating the problem of basis construction in several interesting application domains.

Chapter 6 studies the problem of basis construction in state space planning, in particular in solving optimization problems called Markov decision processes [98]. Here, we will see new challenges present themselves: the set of samples must be dynamically constructed by doing random walks in state spaces. The function to be approximated, called a value function, is not known a priori, but only gradually uncovered through solving a nonlinear equation called the Bellman equation. We will explore both Fourier and wavelet bases in approximately solving Markov decision processes [79].

Chapter 7 explores the application of representation discovery to 3D computer graphics. A crucial application of Fourier and wavelet analysis is compression of multimedia content, such as images, using well-known compression methods such as JPEG and JPEG-2000 [3, 122]. These approaches assume a rigid 2D data structure, and do not extend to graphs with arbitrary topology. We will apply the ideas of Fourier and wavelet analysis on graphs introduced earlier, and show how new compression algorithms may be devised for 3D graphics [59, 76]. A major

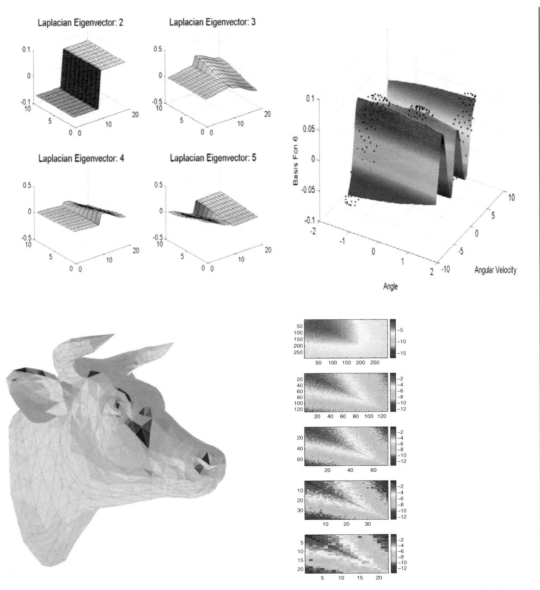

FIGURE 1.4: Top left: Low-frequency eigenvectors of the combinatorial graph Laplacian [26] in a discrete "two-room" spatial environment. Top right: An eigenfunction of the manifold Laplacian [99] in a continuous 2D robotics task, showing samples (dots) and the surface generated by the Nyström interpolation [41]. Bottom left: A *scaling function* from a multiscale diffusion wavelet analysis [30] of a 3D object in computer graphics, capturing the region representing one of the horns (blue region). Bottom right: A diffusion wavelet multiscale representation of a Markov chain, showing the original transition matrix (top), and each successive power of two represented in an increasingly compressed form on a newly generated basis. Matrix entries are shown in the $\log_{10}$ scale.

challenge in computer graphics is scalability, since 3D objects can be very large with millions of vertices. We will explore divide-and-conquer approaches, such as graph partitioning [60], where basis functions are constructed on smaller subgraphs.

Chapter 8 studies the application of basis construction to information retrieval [53] and natural language processing, an area of rapidly growing importance within AI due to the large set of text corpora available on the World Wide Web. We will contrast Fourier and wavelet approaches to topic discovery [18], where text documents are clustered based on finding words that co-occur. Finally, Chapter 9 concludes with a brief discussion of some directions for future work, including recent work in an area called compressed sensing [22], the use of harmonic analysis with richer representations such as logic [13], and the use of group representation theory to construct compact bases over homogeneous spaces.

1.4   BIBLIOGRAPHICAL REMARKS

A detailed description of the history behind the decimal and other number systems can be found on Wikipedia^{TM}. Strang's book [110] (Section 7.3) contains a nice overview of the analysis–synthesis perspective and basis construction. Mallat's text [81] provides a comprehensive overview of Fourier and wavelet analysis in Euclidean spaces. Terras [115] surveys Fourier representations on general algebraic structures, including graphs and groups.

CHAPTER 2

Vector Spaces

In the first part of this book, beginning with this chapter, we review the mathematical foundations of representation theory. The concept of a *basis* is defined, both in the coordinate-dependent finite-dimensional linear algebraic setting [110] as well as the infinite-dimensional setting of inner product spaces [37]. We introduce the *analysis–synthesis* perspective of decomposing a vector into components based on measurements, and then reassembling the vector, which serves as a recurring theme in later chapters. This perspective also helps explain the concept of a dual basis. In the finite-dimensional case, matrix representations depend on the choice of bases over the input (row) and output (column) vector spaces. This explicit dependence of a representation on a choice of basis will turn out to be essential later in the book.

2.1 ANALYSIS–SYNTHESIS FRAMEWORK

The overarching goal of this book is to describe algorithms for constructing *features* that *represent* entities in a novel space different from the original data or state space. The approach is based on a mathematical theory of representation, which we introduce here. The entities being represented will be abstractly viewed as *vectors v* in a vector space *V*. Concretely, in terms of the applications of interest, these entities are value functions in a planning problem [98], training instances in a machine learning problem [16], 3D models in a computer graphics context [76], or text documents in a natural language processing context [18]. All of these entities have a natural representation using some default basis, and the problem addressed in this book is to construct a new representation that better reveals the structure underlying the data or state space.

2.1.1 Approximating 3D Objects in Computer Graphics

To motivate the problem of basis selection, which is discussed at length in this book, we begin with a real-world example from computer graphics [59]. Figure 2.1 illustrates the result of two different choices of bases for approximating a 3D object. Concretely, the problem here is to approximate three *coordinate functions* on a graph of 1107 vertices. On the default unit vector basis, each function is represented using 1107 coefficients—each coefficient specifies the x, y, or z coordinate of a vertex.

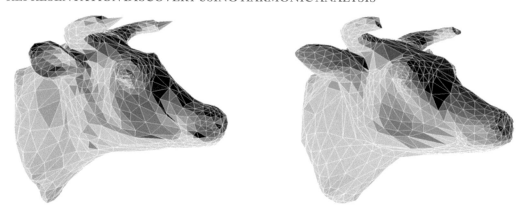

FIGURE 2.1: This figure illustrates the effect of two choices of bases for representing a 3D object in computer graphics. Left: reconstruction with a wavelet basis, described in Chapter 4. Right: the reconstruction using a Fourier basis, described in Chapter 3. Note the wavelet approach excels at reproducing sharp discontinuities like the "horns".

Figure 2.1 contrasts two alternative representations—one using Fourier frequency-based eigenvector bases [26] and the other using multi-resolution wavelet bases [30]—both of which were "discovered" specifically for representing this object by analyzing the *topology* of the object. Using the Fourier basis, each coordinate function can be approximated to an accuracy of 0.004 using just 200 coefficients (the details will be described in Chapter 7), resulting in 80% compression compared to the unit vector basis representation. Even more effective are the wavelet bases, which require overall only 100 numbers to specify the coordinate function, a compression efficiency of 90%.

Of course, the compression efficiency is not the only consideration—there are other issues that need to be considered. The unit vector bases are compact by their very nature, and the Fourier and wavelet bases take up a lot more space. The Fourier and wavelet bases require far fewer coefficients, but the cost of computing each basis needs to be considered (for example, computing k eigenvectors on a graph of N vertices requires $O(kN^2)$ steps [47]). The Fourier bases are less compact than the wavelet basis, but the wavelet bases provide a multi-resolution analysis. Thus, we see that we are trading off issues such as sparsity, compactness, expressiveness, and tractability.

2.1.2   Abstract Fourier Expansion

We now introduce a very useful perspective called the *analysis–synthesis* framework [37, 110, 22], which underlies our approach. Given an object v, a feature-based representation can be viewed as the reconstruction of the object from a set of *measurements*. We will model features in this chapter as a set of *basis vectors* $\phi_i \in V$. Measurements will be abstractly modeled as *linear functionals* $\langle v, \phi \rangle : V \times V \to \mathbb{R}$ (we will define linear functionals precisely later in Section 2.5.2, but for now treat them as mappings from pairs of vectors to real numbers). The process of analysis of an entity v is construed as constructing a set of measurements using the features:

$$v \Rightarrow \{\langle v, \phi_1 \rangle, \ldots, \langle v, \phi_n \rangle\}.$$

The reverse process of *synthesis* can be seen as reconstructing the object from the measurements. In the ideal situation, perfect reconstruction is possible, and the synthesis space uses the same basis vectors as the analysis space. But, for flexibility and computational tractability, it is sometimes beneficial to think of the synthesis space as comprising a set of *dual basis* vectors [110]. The synthesis phase can be modeled as taking a *linear* combination of basis vectors, with the measurements constituting the *weights* associated with the basis vectors or features in the dual space:

$$v = \sum_{i=1}^{\infty} c_i \psi_i = \sum_{i=1}^{\infty} \langle v, \phi_i \rangle \psi_i. \tag{2.1}$$

Here, ψ_i are the synthesis features. In a finite-dimensional vector space, the summation will of course be finite. Often, both the analysis and synthesis bases are *orthogonal*, and in fact, one can be formed from the other. In the more general setting, each set of bases may not be orthogonal, but each basis in the analysis set is *bi-orthogonal* with its corresponding element in the synthesis set. While perfect reconstruction may be possible if all measurements are made, and all synthesis features are used, often the object can be only approximately reconstructed. The challenge of representation discovery is to construct dual bases such that this approximation problem is solved optimally.

Equation (2.1) is sometimes referred to as the *abstract Fourier series* expansion of a vector [37], and will play a central role throughout this book. We now explain the analysis–synthesis perspective in more detail in this chapter, beginning with the concept of a dual basis in finite-dimensional vector spaces, and culminating in the abstract Fourier expansion in a general infinite-dimensional function space.

2.1.3 Issues in Basis Construction and Selection

In this book, we are most often interested in a *partial* basis, where some basis elements are discarded in the Fourier expansion (Equation (2.1)). We have not yet specified how this choice is made. Broadly speaking, there are two approaches. In the *fixed* approach, the subset of basis vectors is selected from the full set, given some a priori fixed ordering on the basis elements such as *smoothness*. In the *adaptive* approach, the subset of basis vectors selected depends on the vector v being approximated [82]. Regardless of which approach is used, let I refer to the set of indices of the selected basis vectors. Then, the *approximate Fourier expansion* is given as

$$v \approx \sum_{i \in I} c_i \psi_i = \sum_{i \in I} \langle v, \phi_i \rangle \psi_i. \tag{2.2}$$

There can be many strategies for selecting the bases out of order. For example, one strategy is to choose the basis vectors (features) that have the largest inner product (measurement values) for the vector v being approximated [82]. That is, the set of basis vectors is resorted for each vector being approximated by the magnitude of the coefficients $c_i = \langle v, \phi_i \rangle$. We will have much more to say about basis selection in this and later chapters, but we define below one key consideration involved in constructing and choosing a basis.

We call a basis $B = \{\phi_1, \ldots, \}$ as *efficient* if for some space of vectors V_B, the approximate Fourier expansion of any $v \in V_B$ involves making the fewest measurements (coefficients), for a desired precision ϵ. More formally, we have

$$||v - \sum_{i \in S_B(\epsilon)} \alpha_i v_i|| \leq \epsilon$$

and furthermore, for any other choice of basis B', the size $|S_{B'}(\epsilon)| > |S_B(\epsilon)|$. Here, $\alpha_i = \langle v, \phi_i \rangle$, and $S_B(\epsilon)$ refers to an index set of basis vectors, chosen either adaptively or using a fixed ordering.

It is important to stress that there are many other objectives that need to be considered, including *compactness*, *sparsity*, and *complexity*, and these may be in conflict. For example, a basis B may be efficient at representing "smooth" functions in a certain class (e.g. smoothness is defined more precisely in Chapter 3), but almost any choice of such a basis will result in a less compact basis than the unit vector basis.

Similarly, we describe algorithms for basis construction on graphs in Chapters 3 and 4. These algorithms result in bases that are well-adapted to approximating functions on graphs with a specific (usually highly irregular) topology. However, the costs of constructing such bases can be expensive, particularly for large graphs. Also, our perspective on basis construction assumes that the goal is to solve some *class* of problems on a given domain, so that the cost of constructing the basis can be amortized. It can be hard to justify the cost of basis construction if the ultimate goal is to solve one specific problem, since in that case, it may turn out that constructing the basis is as expensive as solving the problem in the original (non-efficient) basis.

2.2   DUAL BASES

We begin by explaining the concept of a *dual basis* in finite-dimensional vector spaces. Let us begin with a brief summary of basic linear algebra. A *vector space* is a set of elements called *vectors* that are algebraically closed under addition and multiplication by *scalars* from a *field*. The concept of a vector space is far more general than $\mathbb{R}^n$, the n-dimensional Euclidean space of real numbers: the set of all symmetric matrices, polynomial and trigonometric functions, and so on are all examples of vector spaces.

To illustrate the concept of basis, consider the vector space that is most familiar to us: $\mathbb{R}^3$, or three-dimensional Euclidean space. A fundamental property of a basis is that each vector $v \in \mathbb{R}^3$ can be written uniquely as a *linear* combination of bases. A notational remark: vectors in this book are generally denoted by lower case letters toward the end of the alphabets, such as u, v, w, x, and y. Basis vectors are generally denoted by Greek letters such as ϕ_i or ψ_i. Scalars are denoted by Greek or Roman letters from the beginning of the alphabets, such as α, a, b, and so on:

$$v = \begin{bmatrix} x \\ y \\ z \end{bmatrix} = xu_1 + yu_2 + zu_3 = x\begin{bmatrix} 1 \\ 0 \\ 0 \end{bmatrix} + y\begin{bmatrix} 0 \\ 1 \\ 0 \end{bmatrix} + z\begin{bmatrix} 0 \\ 0 \\ 1 \end{bmatrix}.$$

Bases have two key properties: *completeness*—every vector can be written as a linear combination of basis vectors—and *uniqueness*, the coefficients in the expansion are unique. The uniqueness property, however, is also a weakness, since it provides for no redundancy.

Even for the vector space $\mathbb{R}^n$, there are many choices of bases other than the *unit* or *default* bases such as u_1, u_2, and u_3 above. For example, the *Haar* basis [82] for $\mathbb{R}^4$, which is the earliest example of a *wavelet* basis, is shown below:

$$v_{b_1} = \begin{bmatrix} 1 \\ 1 \\ 1 \\ 1 \end{bmatrix}, \quad v_{b_2} = \begin{bmatrix} 1 \\ 1 \\ -1 \\ -1 \end{bmatrix}, \quad v_{b_3} = \begin{bmatrix} 1 \\ -1 \\ 0 \\ 0 \end{bmatrix}, \quad v_{b_4} = \begin{bmatrix} 0 \\ 0 \\ 1 \\ -1 \end{bmatrix}.$$

Another property of a basis is that if we view the basis vectors as the columns of a *basis matrix*, the resulting matrix represents an invertible linear mapping on the underlying vector space. In other words, the basis matrix is *invertible*. This makes it possible to transform one basis into another. Since this concept is crucial to much of the rest of the book, we explain it in detail in Section 2.3. A mathematical rationale for choosing one basis over another will be discussed in Section 2.6 in the context of *least-squares approximation* [37]. Intuitively, an efficient basis is one where any given vector in the space (or more often, in a subspace) can be expressed using the least number of coefficients.

Let us consider why we might prefer the Haar basis over the unit vector basis. If we consider a constant vector $v = [v_1, \ldots, v_1]^T$ [1], it is clear why the Haar basis is much more

---

[1]In this book, vectors are always interpreted to be column vectors, and will often be shown transposed to save space.

efficient than the unit vector basis. If $n = 10^5$, it takes 10^5 numbers to write out the constant vector in the unit vector basis, but only 1 number in the Haar basis! Clearly, this is a somewhat contrived example, but consider the following less-obvious example:

$$v = \begin{bmatrix} 5 \\ 4.5 \\ -4 \\ -5 \end{bmatrix} = 0.125 \begin{bmatrix} 1 \\ 1 \\ 1 \\ 1 \end{bmatrix} + 4.625 \begin{bmatrix} 1 \\ 1 \\ -1 \\ -1 \end{bmatrix} + 0.25 \begin{bmatrix} 1 \\ -1 \\ 0 \\ 0 \end{bmatrix} + 0.5 \begin{bmatrix} 0 \\ 0 \\ 1 \\ -1 \end{bmatrix}.$$

Here, we have represented the vector $v = [5, 4.5, -4, -5]^T$ in the Haar basis. From the coefficients, it is clear that one basis vector dominates the rest, and we can construct a reasonable approximation of this vector using only its second coefficient in the Haar basis. In other words, we get the approximation

$$v = \begin{bmatrix} 5 \\ 4.5 \\ -4 \\ -5 \end{bmatrix} \approx 4.625 \begin{bmatrix} 1 \\ 1 \\ -1 \\ -1 \end{bmatrix} = \begin{bmatrix} 4.625 \\ 4.625 \\ -4.625 \\ -4.625 \end{bmatrix},$$

where the error in approximation is around 0.8 (measured in terms of the "length" of $\|v - \hat{v}\|_F$ using the Euclidean norm). It might be hard to see this as a significant issue in $\mathbb{R}^4$, but in many of the applications to be discussed later, such as computer graphics and natural language processing, we are dealing with high-dimensional vector spaces such as $\mathbb{R}^{100,000}$ where such differences will be very significant.

More generally, we are interested in approximating vector spaces of *functions*, both in discrete spaces such as graphs where function spaces are still finite-dimensional since they can be viewed as vectors [26], as well as in continuous spaces where function spaces are infinite-dimensional [82, 99]. In these richer settings, the choice of bases will prove to be crucial, and we will explore algorithms for basis construction which dynamically construct a basis during the course of solving an optimization problem.

2.3   LINEAR MAPPINGS AND MATRIX REPRESENTATIONS

Linear mappings $T : V \to W$ transform vectors from one vector space V into another W (where V and W may be the same) [110]. Linear mappings satisfy the following fundamental property:

$$T(a_1 v_1 + a_2 v_2) = a_1 T(v_1) + a_2 T(v_2) \quad \forall v_1, v_2 \in V, a_1, a_2 \in F.$$

We will often use the term *operator* to denote linear mappings in both finite and infinite-dimensional spaces, particularly in contexts where the vector space V being acted on is a space of functions. In a finite-dimensional space, a linear mapping T can be represented by a matrix

M, only after we have decided on a choice of basis for both the input vector space V and the output vector space W. To make this dependence clear, we will often use the notation $[T]_{B_1}^{B_2}$ to denote the matrix representation of a linear mapping T, where the basis B_1 is used for the domain or input vector space V, and the basis B_2 is used for the range or output vector space W (this notation is adapted from [30], and will be used extensively in Chapter 4).

To define the matrix representation of a linear mapping, it is sufficient to specify its output on the basis vectors in the input space, written in the representation of the output space, since every vector in the input space can be written as a linear combination of basis vectors. To make this precise, let $B_1 = \{u_1, \ldots, u_n\}$ be a set of basis vectors for an n-dimensional vector space V, and let $B_2 = \{v_1, \ldots, v_m\}$ be a set of basis vectors for the m-dimensional output space W. The linear mapping T is specified as follows:

$$T(u_1) = u_1', \quad T(u_2) = u_2', \ldots, T(u_n) = u_n' .$$

To construct the matrix representation $[T]_{B_1}^{B_2}$, we must represent each output vector u_i' in terms of the output basis vectors. Thus, we get

$$u_1' = \sum_{i=1}^{m} a_{i1} v_i, \ldots, u_n' = \sum_{i=1}^{m} a_{in} v_i .$$

The matrix representation $[T]_{B_1}^{B_2}$ is then specified by writing the effect of the operator T on each input basis vector in terms of the output basis vectors as the corresponding column of the matrix:

$$[T]_{B_1}^{B_2} = \begin{bmatrix} a_{11} & \cdots & a_{1n} \\ a_{21} & \cdots & a_{2n} \\ & \vdots & \\ a_{m1} & \cdots & a_{mn} \end{bmatrix}.$$

To make this concrete, consider the identity mapping $I : \mathbb{R}^4 \to \mathbb{R}^4$, where the input space is the Haar basis and the output space is the unit basis. That is, I is defined as follows:

$$I(v_{h_1}) = v_{h_1}, \quad I(v_{h_2}) = v_{h_2}, \quad I(v_{h_3}) = v_{h_3}, \quad I(v_{h_4}) = v_{h_4}.$$

To define the matrix of the identity mapping, we now have to express each of the Haar basis vectors in terms of the unit vectors $u_1 = [1, 0, 0, 0]^T$, $u_2 = [0, 1, 0, 0]^T$, $u_3 = [0, 0, 1, 0]^T$, and $u_4 = [0, 0, 0, 1]^T$. In this case, the matrix is just the Haar basis vectors written out as each column:

$$[I]_H^U = \begin{bmatrix} 1 & 1 & 1 & 0 \\ 1 & 1 & -1 & 0 \\ 1 & -1 & 0 & 1 \\ 1 & -1 & 0 & -1 \end{bmatrix}.$$

We will use a convenient abbreviation for this type of basis change matrix, namely $[I]_H^U = [H]_U$ will denote the Haar basis vectors written in terms of the unit vector basis U. More generally, $[I]_{B'}^B = [B']_B$ represents the basis vectors B' in terms of the basis B. Note that if the input and output bases are the same, $[I]_B^B = I$, the identity matrix. By inverting the basis change matrix, we can set up an isomorphism between two bases. For example, if we represent a vector in the unit basis, and wish to determine its representation in the Haar basis, we need to invert the above matrix:

$$[U]_H = [I]_U^H = ([I]_H^U)^{-1}$$

$$= \begin{bmatrix} 1 & 1 & 1 & 0 \\ 1 & 1 & -1 & 0 \\ 1 & -1 & 0 & 1 \\ 1 & -1 & 0 & -1 \end{bmatrix}^{-1} = \begin{bmatrix} 0.25 & 0.25 & 0.25 & 0.25 \\ 0.25 & 0.25 & -0.25 & -0.25 \\ 0.5 & -0.5 & 0 & 0 \\ 0 & 0 & 0.5 & -0.5 \end{bmatrix}. \quad (2.3)$$

To check that this is correct, note that

$$0.25v_{b1} + 0.25v_{b2} + 0.25v_{b3} + 0v_{b4} = [1, 0, 0, 0]^T = u_1.$$

Let us denote $[v]_B$ to mean the representation of the vector v in the basis B. If we want to determine the representation $[v]_{B'}$ in some other basis B', we need to use the basis change matrix $[B']_B$, which represents the new basis B' in terms of the old basis B:

$$[v]_{B'} = [I]_B^{B'} [v]_B = [B]_{B'} [v]_B.$$

In the last section, we gave two representations of the same vector, one in the unit basis, and the other in the Haar basis:

$$[v]_U = [5, 4.5, -4, -5]^T, \quad [v]_H = [0.125, 4.625, 0.25, 0.5]^T.$$

How did we determine the Haar representation $[v]_H$? Here, we need to invert the basis change matrix giving us

$$[v]_H = [I]_U^H [v]_U = [U]_H [v]_U = ([I]_H^U)^{-1} [v]_U = \begin{bmatrix} 0.125 \\ 4.625 \\ 0.25 \\ 0.5 \end{bmatrix}.$$

As this simple example suggests, changing a basis depends crucially on the properties of the basis change matrix. The Haar basis change matrix has some desirable properties, such as decomposability, which makes it very efficient to apply the inverse to determine the new representation [110]. Since the Haar basis change matrix $[I]_H^U$ is *orthogonal*, its inverse is just its transpose (we have to correct for the fact that its columns are not unit length, but that is a trivial matter). A more subtle property of the Haar matrix is that it decomposes into the product of smaller matrices, which makes its application much more efficient. To understand matrix decomposition more generally, we turn now to explain the notion of an *invariant* subspace of a vector space.

2.4   INVARIANT SUBSPACES

The approach to representation discovery described in this book can abstractly be characterized as determining the *invariant subspaces* of a vector space under some operator T, and building basis functions that span these subspaces. A subspace χ of a vector space V is invariant under a linear mapping T if for each vector $w \in \chi$, the result $Tw \in \chi$.

Invariant subspaces are useful to identify since they enable representing linear mappings using *irreducible representations*. In this chapter, the notion of invariance is tied to the action of a linear operator on a subspace. A key theorem regarding invariant subspaces is worth stating and proving formally [109].

Theorem 2.1. *Let χ be an invariant subspace of T, and let the columns of matrix X form a basis for χ. Then, there is a unique matrix L such that*

$$TX = XL.$$

In other words, the matrix L is the representation of T on the subspace χ with respect to the basis X. It is often useful to refer to the restriction *of an operator T on a subspace χ as $T|_\chi$. In terms of the notation introduced earlier, $L = [T|_\chi]_X^X$.*

The proof is straightforward (see [109]). Since χ is an invariant subspace, for any vector $x_i \in \chi$, $Tx_i \in \chi$, and consequently can be expressed as a linear combination of the columns in X. That is, $Tx_i = Xl_i$, where l_i is the unique set of coefficients. The matrix $L = [l_1, \ldots, l_n]$.

Often, we can determine a set of invariant subspaces such that every vector in a vector space can be written as the *direct sum* of vectors from each subspace. That is, given any $v \in V$, we can write it as

$$v = w_1 + w_2 + \cdots + w_k,$$

where each $w_i \in W_i$, an invariant subspace of an operator T. We will use the following notation for the direct sum decomposition of V:

$$V = W_1 \oplus W_2 \oplus \cdots \oplus W_k.$$

How do we find an invariant subspace? There are several interesting alternatives, which we will explore more thoroughly in the remainder of the book. We will summarize below a few main avenues for constructing invariant subspaces.

2.4.1   Dual Bases and Direct Sum Decompositions

Linear algebra can be viewed as the study of matrix representations under different choices of bases for the input (row) and output (column) spaces. One guide to developing interesting matrix decompositions is to look for ways of representing a matrix as a linear sum of simpler *rank-one* matrices, which are essentially just the *outer product* of two vectors [47].

More formally, the *rank* of a matrix A is the number of independent columns (or rows). The basis change matrices above, such as $[I]_H^U$ were *full-rank* matrices because the rank was equal to the number of columns. Such matrices are invertible, and we can use this property to define the notion of a *dual* basis. Given an invertible basis change matrix $[I]_{B_1}^{B_2}$, we know that

$$([I]_{B_1}^{B_2})^{-1}[I]_{B_1}^{B_2} = \begin{bmatrix} u_1^T \\ u_2^T \\ \vdots \\ u_n^T \end{bmatrix} \begin{bmatrix} v_1 & \cdots & v_n \end{bmatrix} = I,$$

where u_i^T indicates the ith row of the basis change matrix from basis B_2 into basis B_1, and v_i represents the columns of the basis change matrix in the reverse direction from B_1 to B_2. Since the inner product of u_i and v_i gives us the identity matrix, these vectors must necessarily be *orthogonal* to one another, and hence the vectors u_i represent the *dual basis* to the original basis vectors v_i. For example, the dual basis to the Haar basis are the rows of the inverse of the basis change matrix, which was given in Equation (2.3), and written below as column vectors (where

v'_{b_i} represents the dual basis to the ith Haar basis vector v_{b_i}):

$$v'_{b_1} = \begin{bmatrix} 0.25 \\ 0.25 \\ 0.25 \\ 0.25 \end{bmatrix}, \quad v'_{b_2} = \begin{bmatrix} 0.25 \\ 0.25 \\ -0.25 \\ -0.25 \end{bmatrix}, \quad v'_{b_3} = \begin{bmatrix} 0.5 \\ -0.5 \\ 0 \\ 0 \end{bmatrix}, \quad v'_{b_4} = \begin{bmatrix} 0 \\ 0 \\ 0.5 \\ -0.5 \end{bmatrix}.$$

Note the dual basis vectors are just the scaled versions of the original basis vectors. This is no accident: if the original change of basis matrix is orthogonal, the dual basis matrix is just its (scaled) transpose. We can now combine the Haar basis with its dual basis to construct a set of invariant subspaces that results in a direct sum decomposition of the original vector space V. Let us illustrate this decomposition for the vector space $\mathbb{R}^4$. Note that we can write a vector in the unit basis as

$$[w]_U = ([I]_H^U)([I]_H^U)^{-1}[w]_U = \sum_i^n v_{b_i}(v'_{b_i})^T[w]_U.$$

Taking the same vector as before, $[v]_U = [5, 4.5, -4, -5]^T$, we can decompose this vector as the sum of four vectors, each drawn from one of the invariant subspaces resulting from the rank-one matrices $v_{b_i}v'_{b_i}$, each of which is produced by taking the outer product of the Haar basis vectors with the corresponding dual basis vectors:

$$[v]_U = \begin{bmatrix} 5 \\ 4.5 \\ -4 \\ -5 \end{bmatrix} = \begin{bmatrix} 0.125 \\ 0.125 \\ 0.125 \\ 0.125 \end{bmatrix} + \begin{bmatrix} 4.625 \\ 4.625 \\ -4.625 \\ -4.625 \end{bmatrix} + \begin{bmatrix} 0.25 \\ -0.25 \\ 0 \\ 0 \end{bmatrix} + \begin{bmatrix} 0 \\ 0 \\ 0.5 \\ -0.5 \end{bmatrix}.$$

Note that each of these subspaces can now be given a meaning. For example, the first subspace represents the "mean" component of the vector because $0.125 = \frac{5+4.5-4-5}{4} = 0.125$. Similarly, the second subspace represents the difference between the first two components and the second two components $4.625 = \frac{5+4.5-(-4+-5)}{4}$. We can write the matrix representation of these two subspaces as follows:

$$v_{b1}(v'_{b1})^T = \begin{bmatrix} 0.25 & 0.25 & 0.25 & 0.25 \\ 0.25 & 0.25 & 0.25 & 0.25 \\ 0.25 & 0.25 & 0.25 & 0.25 \\ 0.25 & 0.25 & 0.25 & 0.25 \end{bmatrix}, \quad v_{b2}(v'_{b2})^T = \begin{bmatrix} 0.25 & 0.25 & -0.25 & -0.25 \\ 0.25 & 0.25 & -0.25 & -0.25 \\ -0.25 & -0.25 & 0.25 & 0.25 \\ -0.25 & -0.25 & 0.25 & 0.25 \end{bmatrix}$$

and their highly regular structure makes it possible to devise fast recursive basis change algorithms for wavelet bases such as the Haar basis.

2.4.2 QR Decomposition and Gram–Schmidt Orthogonalization

We now investigate more general methods for constructing invariant subspaces for general matrices, where the original matrix may not be invertible, or the columns of the matrix may not be independent. In constructing new representations of a matrix A, it is possible to change the output basis, the input basis, or both. We begin with a well-known method that only changes the output basis called QR-decomposition [47, 109].

If the columns of a matrix A are not orthogonal, a standard method called Gram–Schmidt orthogonalization [110] can be applied that results in the construction of an invariant subspace Q, with respect to which the matrix A can be represented as a triangular matrix R. In other words, using the notation we have developed previously, we can write $A = QR$, or in other words, $R = [A]_U^Q$.

Note that we have only modified the output basis of the matrix A, and hence its input basis remains the same (by default, assumed here to be the unit basis). A simple (but not necessarily the most efficient) algorithm for Gram–Schmidt orthogonalization is as follows:

1. Define the first basis vector $q_1 = a_1$, the first column vector of A.

2. Determine the component of the second basis vector that lies in a direction orthogonal to the first basis vector:[2]

$$q_2 = a_2 - \frac{a_2^T q_1}{\|q_1\|}.$$

3. Repeat the previous step for each column vector of A.

4. Normalize each vector q_i by dividing it by its length $\|q_i\|$.

2.4.3 Eigenspace Decomposition

We now describe ways of transforming both the input and output bases of a matrix. One of the most useful ways of finding invariant subspaces is through *eigenspace* analysis. Eigenspaces are invariant subspaces that are one-dimensional, and associated with scalar eigenvalues [110]. More formally, x is an eigenvector associated with the eigenvalue λ when

$$Ax = \lambda x = x\lambda.$$

Note the similarity with Theorem 2.1: the space spanned by the eigenvector x is an invariant space. Furthermore, λ is the representation of A on the space spanned by x. We can rewrite the above equation using our basis notation as

$$[A|_\chi]_x^x = \lambda,$$

---

[2]We discuss orthogonality and projections in the more general case in Section 2.5.2.

where χ is the subspace spanned by the eigenvector x. If the vector space V is defined over the field of real numbers $\mathbb{R}$, it is easy to show that not all matrices have eigenvectors. Consider the basis change transformation produced by rotating the coordinates of $\mathbb{R}^2$ by an angle θ. The matrix corresponding to this basis transformation is defined as

$$R = \begin{bmatrix} \cos(\theta) & \sin(\theta) \\ -\sin(\theta) & \cos(\theta) \end{bmatrix}.$$

However, over the field of complex numbers $\mathbb{C}$, such rotation matrices do admit eigenvectors. Since we often need to consider the field of complex numbers, a brief review is in order. A complex number $z = (a, b)$ is defined by a pair of real numbers $a \in \mathbb{R}$ and $b \in \mathbb{R}$, with the rules of addition and multiplication defined as follows:

$$(a, b) + (c, d) = (a + c, b + d) \quad (a, b) \times (c, d) = (ac - bd, ad + bc).$$

Complex numbers are often denoted as $a + ib$, where $i^2 = -1$. The rule for multiplication is easily derived from the product $(a + ib)(c + id)$. Returning to the problem above of determining the eigenvectors of a rotation matrix, let us consider the simple case of rotation by $\theta = 90°$. The corresponding eigenvalues and eigenvectors in this case are

$$\begin{bmatrix} 0 & 1 \\ -1 & 0 \end{bmatrix} \begin{bmatrix} 1 \\ i \end{bmatrix} = i \begin{bmatrix} 1 \\ i \end{bmatrix}, \quad \begin{bmatrix} 0 & 1 \\ -1 & 0 \end{bmatrix} \begin{bmatrix} 1 \\ -i \end{bmatrix} = -i \begin{bmatrix} 1 \\ -i \end{bmatrix}.$$

It can be shown that any linear transformation admits at least one eigenvalue and eigenvector over $\mathbb{C}$ [4]. Let us consider linear transformations whose matrix representations admit a *complete* set of eigenvectors. Such matrices are termed *diagonalizable* because of the following property:

$$A[x_1, \ldots, x_n] = [\lambda_1 x_1, \ldots, \lambda_n x_n] \Rightarrow AS = S\Lambda \Rightarrow \Lambda = S^{-1} A S,$$

where S is the matrix of eigenvectors, and Λ is a diagonal matrix of eigenvalues. Note that we have now transformed both the output *and* input basis of the matrix A to the basis represented by the eigenvectors. For an $n \times n$ matrix to be diagonalizable, the matrix must admit n different eigenvalues, since each eigenvalue leads to (at least) one different eigenvector.

For special classes of matrices, such as *symmetric* matrices, where $A^T = A$, all eigenvalues are *real*, and there is a complete set of *orthonormal* eigenvectors. This result is so important for the remainder of the book that it is worth stating it formally as a theorem [110].

Theorem 2.2. *Any $n \times n$ symmetric matrix A can be diagonalized as $A = Q\Lambda Q^T$, where Q is an orthonormal matrix whose columns are the (real) eigenvectors, and Λ is a diagonal matrix of n (real)* *eigenvalues:*

$$A = Q \Lambda Q^T = Q \Lambda Q^{-1}, \text{ where } Q^{-1} = Q^T.$$

The orthonormality of the eigenvector matrix Q means that each column q_i is orthogonal to all other columns $q_i^T q_j = 0$, $i \neq j$, and furthermore, $q_i^T q_i = 1$. One of the major applications of this theorem will be in constructing a basis for functions on an undirected graph $G = (V, E)$. The adjacency matrix of an (unweighted) undirected graph G is given by $A(i, j) = 1$ if and only if $(i, j) \in E$ is an edge from vertex $i \in V$ to $j \in V$. Adjacency matrices are clearly diagonalizable since they are symmetric, and the resulting orthonormal basis of eigenvectors can be used to represent any function $f : V \to \mathbb{R}$ on the graph G.

Positive-definite matrices are a special class of symmetric matrices where all the eigenvalues are not just real, but also positive. A matrix A is positive-definite if for any nonzero vector x, $x^T A x > 0$. *Positive semi-definite* matrices allow for eigenvalues to be 0, so that $x^T A x \geq 0$. In Chapter 3, we will investigate an important class of positive semi-definite matrices called Laplacian matrices [26].

2.4.4   Singular Value Decomposition

Arguably, the single most celebrated result in linear algebra is the singular value decomposition (SVD) [47], which shows that any matrix representation of a finite-dimensional linear transformation $A : V \to W$ can be diagonalized. This result depends on the construction of not one, but two sets of bases that when combined with each other, lead to a new way of constructing invariant subspaces.

The key idea, as before, is to express the result of the linear transformation Au_i on the input basis vectors as a linear combination of output basis vectors v_i. However, what is unique about SVD is that the input and output bases are not arbitrary, but selected in such a way that $Au_i = \sigma_i v_i$. In other words, the result of applying the linear transformation to u_i is a vector that does not point in the direction of u_i (which would give rise to eigenvalues), but rather in the direction of v_i, the corresponding output basis vector. Using these input and output basis vectors, any $m \times n$ matrix A can be diagonalized as

$$A = U \Sigma V^T, \tag{2.4}$$

where U and V are orthonormal matrices whose columns are the left and right *singular vectors*, respectively, so that $U^T U = I$ and $V^T V = I$. Σ is a diagonal matrix of *singular values* (not to be confused with eigenvalues). Here, we assume $m \geq n$.

There are two types of SVD decompositions: in the regular SVD, U is of size $m \times m$, Σ is of size $m \times n$, and V is of size $n \times n$. This decomposition can be highly wasteful in problems where the matrix A is "skinny" (that is, $m \gg n$). If A is a set of 100 points in $\mathbb{R}^2$, in the regular

SVD decomposition, Σ is of size 100×2 with only two nonzero entries $\Sigma(1, 1)$ and $\Sigma(2, 2)$. A much better alternative in such cases is the so-called *thin-SVD* [47], where U is a matrix of size $m \times n$, Σ is a diagonal matrix of size $n \times n$, and V is a matrix of size $n \times n$.

One of the most important applications of SVD is in constructing *low-rank* approximations of a matrix A_k, which can be explained by first rewriting the SVD decomposition in a manner that makes explicit the underlying invariant subspaces:

$$A = \sum_{i=1}^{r} \sigma_i u_i v_i^T ,$$

where r is the rank of A. Using this rewritten form of SVD decomposition, the *optimal* rank-k approximation of the matrix A can be expressed as

$$A \approx A_k = \sum_{i=1}^{k} \sigma_i u_i v_i^T.$$

In Chapter 5, we will develop an interesting application of SVD to the problem of constructing a *Kronecker* decomposition of a matrix A into smaller matrices B and C such that $A \approx B \otimes C$ [120].

2.5   BASES IN INFINITE-DIMENSIONAL SPACES

We now turn to develop a *coordinate-free* view of bases, where we will not write down explicit matrix representations of the linear transformations (or operators). This viewpoint is essential in the infinite-dimensional setting of function spaces on $\mathbb{R}$ and $\mathbb{C}$, but many of these ideas turn out to be very helpful even in finite-dimensional function spaces on graphs.

We introduce the notion of norm in a vector space, and focus on a specific type of norm defined by the inner product, which plays a central role in approximation theory discussed in the following section. Hilbert spaces are a natural setting for formalizing a theory of approximation, since they provide a generalized coordinate-independent notion of projection and orthogonality.

2.5.1   Normed spaces

Intuitively, a norm is an abstraction of *length*. A *normed space* is a (possibly infinite-dimensional) vector space V along with a real-valued function, called the norm $\|x\| : V \to \mathcal{R}$. The norm should satisfy some properties:

- $\|x\| \geq 0$;

- $\|x + y\| \leq \|x\| + \|y\|$ (triangle inequality);

- $\|\alpha x\| = |\alpha| \|x\|$.

For example, if $V = \mathbb{R}^n$, the set of all n-dimensional real numbers, one possible norm is $\|x\|_2 = \sqrt{x_1^2 + \cdots + x_n^2}$. This is usually referred to as the L_2 norm. More generally, the L_p norm is defined as $\|x\|_p = (|x_1|^p + \cdots + |x_n|^p)^{\frac{1}{p}}$. One important case is when $p = \infty$, which is referred to as the *max-norm* in which case $\|x\|_\infty = \max_i |x_i|$. Another example of a norm in an infinite-dimensional case is when the vector space V is the set of all continuous functions $C[a, b]$ on the real interval $[a, b]$ where $\|f\| = \max_{a \leq x \leq b} |f(x)|$.

We will often need to use norms on matrices as well. The *Frobenius* norm $\|A\|_F$ of a matrix A is defined as $\|A\|_F = \sqrt{\sum_{i,j} a_{ij}^2}$. That is, it is the L_2 norm of the matrix viewed as a vector, constructed by stacking the columns of the matrix on top of each other. The *spectral* norm is defined as

$$\|A\|_2 = \max_{x \in \mathbb{R}^n} \frac{\|Ax\|_2}{\|x\|_2}.$$

If a matrix is symmetric, clearly $\|A\|_2 = |\lambda_{\max}|$. For a positive-definite matrix, $\|A\|_2 = \lambda_{\max}$.

2.5.2 Inner Product Spaces

Hilbert spaces are vector spaces equipped with a special type of norm defined by the inner product. The concept of inner products is fundamental, and generalizes the notion of orthogonality in finite-dimensional spaces. Inner product spaces are valuable in understanding approximation methods, including those used in machine learning. Many forms of learning can be viewed as projecting the original data onto a lower dimensional subspace of an inner product space [37]. To capture the concept of orthogonality in a general (infinite-dimensional) vector space, we need to define a special kind of norm. An *inner product space* is a normed space V, where the norm is defined by an inner product $\langle x, y \rangle : V \times V \to \mathbb{R}$.[3] Intuitively, the inner product $\langle x, y \rangle$ measures the *similarity* between two objects x and y. Formally, inner products satisfy the following properties:

- $\langle x, y \rangle = \langle y, x \rangle$ (symmetry);[4]
- $\langle x + y, z \rangle = \langle x, z \rangle + \langle y, z \rangle$ (distributive law);
- $\langle \lambda x, y \rangle = \lambda \langle x, y \rangle$;
- $\langle x, x \rangle \geq 0$ and equality holds iff $x = \theta$ (null element).

The vector space $\mathbb{R}^n$ is an inner product space where the inner product of x and y is defined as $\sum_{i=1}^n x_i y_i$. More generally, the vector space of all *square-integrable* functions $\mathbb{L}^2(\mathbb{R})$

---

[3]This definition naturally extends to the field of complex numbers $\langle x, y \rangle : V \times V \to \mathbb{C}$.

[4]Over complex numbers, symmetry holds only in a conjugate sense, so that $\langle x, y \rangle = \overline{\langle y, x \rangle}$.

on the real interval $[a, b]$ is an inner product space, where $\langle f, g \rangle = \int_a^b f(x)g(x)dx$. If we consider the space of complex-valued functions over $\mathbb{R}$, we have $\langle f, g \rangle = \int_{-\infty}^{\infty} f(x)\overline{g(x)}dx$, where $\overline{g(x)}$ is the complex conjugate of $g(x)$.

The Cauchy Schwartz inequality holds in any inner product space, and is considered the most important inequality in all of mathematics:

$$|\langle x, y \rangle| \leq \|x\| \|y\| \text{ where equality holds if and only if } x = \lambda y.$$

The two vectors x and y are said to be *orthogonal* if $\langle x, y \rangle = 0$, which is written as $x \perp y$. A vector x is orthogonal to a set S, written $x \perp S$ if $x \perp y$ for all $y \in S$. The generalized *Pythagorean theorem* can now be stated: if $x \perp y$ then $\|x + y\|^2 = \|x\|^2 + \|y\|^2$.

2.5.3 Banach and Hilbert Spaces

We now introduce formally the concept of a *complete* vector space, which is useful in the analysis of convergence of many approximation methods. A sequence of vectors x_n is a *Cauchy* sequence if $\|x_n - x_m\| \to 0$ as $n, m \to \infty$, that is, given any $\epsilon > 0$, there exists an N such that for all $m, n > N$, $\|x_n - x_m\| < \epsilon$. A sequence of vectors x_n is said to *converge* to x if the sequence $\|x - x_n\|$ converges to 0. In a normed space, every convergent sequence is a Cauchy sequence. If $x_n \to x$, then

$$\|x_n - x_m\| = \|x_n - x + x - x_m\| \leq \|x_n - x\| + \|x - x_m\| \to 0.$$

In general, Cauchy sequences need not converge (e.g., consider a sequence of "ramp" functions converging to a step function). A normed vector space X is *complete* if every Cauchy sequence in X has a limit in X. Such complete spaces are called *Banach* spaces. A complete inner product space is called a *Hilbert* space [37].

2.6   PROJECTIONS

The concept of projections will play a central role in this book: given a basis B, spanning a subspace, we are often interested in the closest vector $\hat{x}$ to x that lies in the subspace spanned by the basis B. We begin by reviewing the concept of projections onto the column space of a matrix, before generalizing the concept of projections to an arbitrary Hilbert space. In Section 6.2.1, we will see how to formulate value function approximation in Markov decision processes as approximation in a Hilbert space.

2.6.1   Projections onto Finite-Dimensional Spaces

Given a basis set of linearly independent vectors $a_1, \ldots, a_n$ in R^m, we can find the combination $\hat{b} = \hat{x}_1 a_1 + \cdots + \hat{x}_n a_n$ that is closest to a given vector b that lies outside the subspace spanned by the basis. Let us treat the vectors $a_1, \ldots, a_n$ as forming a matrix A. Since the "error" vector $b - A\hat{x}$ must be *orthogonal* to the space spanned by the basis vectors a_i, we get

$$a_1^T(b - A\hat{x}) = 0, \ldots, a_n^T(b - A\hat{x}) = 0.$$

These n equations can be summarized as $A^T(b - A\hat{x}) = 0$, where A is an m, n rectangular matrix. Solving this equation gives us $\hat{x} = (A^TA)^{-1}A^Tb$, or $\hat{b} = A(A^TA)^{-1}A^Tb = pb$, where the *projection matrix* is given by

$$p = A(A^TA)^{-1}A^T.$$

Note that in terms of the pseudo inverse of the matrix, we can write this as $p = AA^\dagger$, where $A^\dagger = (A^TA)^{-1}A^T$.

2.6.2   Projections in Infinite-Dimensional Hilbert Spaces

A fundamental assumption made in the derivation of the solution to the problem of projecting onto the column space of the matrix is that the error vector $b - A\hat{x}$ is orthogonal to the basis vectors spanning the column space. This concept can be generalized to an arbitrary Hilbert space, where it is worth stating as a theorem [37].

Theorem 2.3. *Let u be a vector in a Hilbert space V, and let M be a closed subspace of V. Then, the closest vector $\hat{u} \in M$ to u is one that minimizes the distance $\|v - u\|$ over all other vectors $v \in M$, and also the unique vector such that $u - \hat{u}$ is orthogonal to M.*

Let us now generalize the concept of projection to a general Hilbert space. Let u be a vector in a Hilbert space. We want to "approximate" u by finding a vector $\hat{u}$ in a subspace M that is "closest" in norm to u. Since $\hat{u} \in M$, it can be expressed as a linear weighted sum of orthonormal basis vectors $\hat{u} = \sum_i \alpha_i \phi_i$, where $\{\phi_1, \phi_2, \ldots\}$ form an orthonormal basis set for the subspace M. Since M is a Hilbert space by itself, the "error" vector $u - \hat{u}$ must be orthogonal to every element of the subspace M. In particular, $\langle u - \hat{u}, \phi_j \rangle = 0$. This gives us the following identities:

$$\langle u - \hat{u}, \phi_j \rangle = 0 \quad \Rightarrow \quad \langle (u - \sum_i \alpha_i \phi_i), \phi_j \rangle = 0$$

$$\langle u, \phi_j \rangle = \sum_i \alpha_i \langle \phi_i, \phi_j \rangle$$

$$\text{But } \langle \phi_i, \phi_j \rangle = \delta_{ij} \quad \Rightarrow \quad \alpha_i = \langle u, \phi_i \rangle.$$

This gives the solution, which is sometimes referred to as the *abstract Fourier expansion*:

$$\hat{u} = \sum_i \langle u, \phi_i \rangle \phi_i. \tag{2.5}$$

Generally speaking, the problem of reconstructing an unknown vector from a series of *measurements* represented by the inner products $\langle u, \phi_i \rangle$ is of wide interest in a number of areas, including machine learning, optimization, and signal processing. Here, we are assuming that there is a single basis represented by the vectors ϕ_i. It is possible to generalize the Fourier expansion to use a *bi-orthogonal basis*, specified as

$$\hat{u} = \sum_i \langle u, \phi_i \rangle \psi_i, \tag{2.6}$$

where ϕ_i represents the *analysis* basis space and ψ_i represents the *synthesis* basis space.

2.6.3   Reproducing Kernel Hilbert Spaces

We conclude this chapter with a discussion of a special class of Hilbert spaces, called a *Reproducing Kernel Hilbert Space* (RKHS) [104]. It can be shown that the diffusion operators on a graph, such as the graph Laplacian described in Chapter 3, generate a RKHS. A Hilbert space of functions H on a set T is said to be a RKHS if there exists a kernel function $K : T \times T \to \mathbb{R}$ having the following properties: (i) All functions $k(., t) \in H$ for each $t \in T$. (ii) For any function $f \in H$ and $x \in T$, the kernel serves as the *representer* of evaluation of the function on the element x, namely $f(x) = \langle f, k(., x) \rangle$. Such a function $k(x, y)$ is called a *reproducing kernel* because of the above property, and because of the following property: $k(x, y) = \langle k(x, .), k(., y) \rangle$.

Note that $k(x, y)$ must be a symmetric function on $T \times T$, and furthermore, $k(., .)$ is also positive semi-definite. That is, for all finite sequences on the domain T of length n, $\sum_{i,j=1}^{n} \alpha_i \alpha_j k(x_i, x_j) \geq 0$. As we will see in Chapter 3, the graph Laplacian defines a positive semi-definite matrix.

Kernels induce *Gram* matrices on a given set of samples. In Chapter 5, we will see how properties of kernel-induced Gram matrices enable extending basis functions computed on a set of samples to new points. Gram matrices also generalize the concept of a projection matrix. The generalized normal equation is

$$\langle u, \phi_j \rangle = \sum_i \alpha_i \langle \phi_i, \phi_j \rangle.$$

In matrix form, this can be written as $G(\phi_1, \ldots, \phi_n)\alpha = \beta$, where $\alpha = (\alpha_1, \ldots, \alpha_n)^T$ and $\beta = (\langle u, \phi_1 \rangle, \langle u, \phi_2 \rangle, \ldots, \langle u, \phi_n \rangle)^T$. The *Gram matrix* (which is symmetric) is defined as

$$G(\phi_1, \ldots, \phi_n) = \begin{bmatrix} \langle \phi_1, \phi_1 \rangle & \langle \phi_2, \phi_1 \rangle & \cdots & \langle \phi_n, \phi_1 \rangle \\ \langle \phi_1, \phi_2 \rangle & \langle \phi_2, \phi_2 \rangle & \cdots & \langle \phi_n, \phi_2 \rangle \\ \cdots & & & \\ \langle \phi_1, \phi_n \rangle & \langle \phi_2, \phi_n \rangle & \cdots & \langle \phi_n, \phi_n \rangle \end{bmatrix}.$$

2.7   BIBLIOGRAPHICAL REMARKS

Strang [110] provides a highly readable introduction to basic coordinate-dependent finite-dimensional linear algebra. A more advanced introduction to coordinate-free linear algebra emphasizing the operator viewpoint is given in [4]. Deutsch [37] provides a rigorous overview of inner product spaces, and applications to approximation theory. A comprehensive reference for matrix computation is [47]. Theorem 2.1 is from Stewart and Sun [109], which is a detailed study of matrix perturbation theory. Gram matrices play a central role in machine learning in the area of *kernel methods* [104].

CHAPTER 3

Fourier Bases on Graphs

In the previous chapter, we described the construction of invariant subspaces associated with the eigenspaces of a linear mapping on a vector space. We introduced the concept of abstract Fourier analysis corresponding to projections of functions onto basis vectors spanning these invariant subspaces. In the analysis phase, linear functionals provide "measurements" of a given function, and in the synthesis phase, these measurements are then combined with the dual bases to reconstitute the original function.

In this chapter, we explore a specific type of Fourier analysis on (undirected and directed) graphs using a linear operator called the *graph Laplacian* [26]. We will show how Fourier analysis on graphs corresponds to diagonalizing the graph Laplacian and using its eigenspaces as invariant subspaces. The graph Laplacian has found applications in a number of areas in machine learning, ranging from dimensionality reduction [9], clustering [88], segmentation of images in computer vision [106], and the compression of 3D objects in computer graphics [59] to the solution of Markov decision processes [74]. Leaving aside these specific applications to later chapters, we focus specifically on the properties of the graph Laplacian. The graph Laplacian is also closely connected to random walks on a graph [26], and we briefly discuss the connection to reversible Markov chains and the Perron–Fröbenius theorem. We summarize some important spectral properties of the graph Laplacian, which have found numerous applications including graph partitioning [44].

3.1 ANALYSIS–SYNTHESIS PERSPECTIVE REVISITED

We begin by instantiating the general analysis–synthesis perspective in the context of Fourier analysis on graphs. Let $G = (V, E, W)$ represent an undirected graph on $|V| = n$ nodes, where $(u, v) \in E$ is an edge from vertex u to v. Edges are all assumed to have a weight associated with them, specified by the W matrix. For now, we assume $W(u, v) = W(v, u)$, but we will relax this assumption later. We define $u \sim v$ to mean an (undirected) edge between u and v, and the degree of u to be $d(u) = \sum_{u \sim v} w(u, v)$. D will denote the diagonal matrix defined by $D_{uu} = d(u)$, and W the matrix defined by $W_{uv} = w(u, v) = w(v, u)$.

We consider the Hilbert space of functions on a graph, where each function $f : V \to \mathbb{R}$. The inner product between f and g is specified as[1]

$$\langle f, g \rangle = \sum_{v \in V} f(v) g(v).$$

We now formalize the notion of a *smooth* function on a graph. The $\mathbb{L}^2$-norm of a function on G is

$$\|f\|_2^2 = \sum_{v \in V} |f(v)|^2 d(v).$$

The gradient of a function is $\nabla f(i, j) = w(i, j)(f(i) - f(j))$ if there is an edge e connecting i to j, 0 otherwise. The smoothness of a function on a graph can be measured by the *Sobolev norm* [78]:

$$\|f\|_{\mathcal{H}^2}^2 = \|f\|_2^2 + \|\nabla f\|_2^2 = \sum_{v \in V} |f(v)|^2 d(v) + \sum_{u \sim v} |f(u) - f(v)|^2 w(u, v). \qquad (3.1)$$

The first term in this norm controls the size (in terms of $\mathbb{L}^2$-norm) for the function f, and the second term controls the size of the gradient. The smaller $\|f\|_{\mathcal{H}^2}$, the smoother is f. In the applications to be considered later, the functions of interest have small $\mathcal{H}^2$ norms, except at a few points, where the gradient may be large.

As discussed in the previous chapter, the approximation of functions on a vector space of dimension $\mathbb{R}^{|V|}$ is facilitated by defining a basis (or a dual basis), where the analysis basis constructs measurements of the function, and the synthesis basis reconstructs the function from the measurements. For simplicity, let us assume an orthonormal basis $(e_1, \ldots, e_{|V|})$ for the space $\mathbb{R}^{|V|}$. For a fixed precision ϵ, a function f can be approximated as

$$\|f - \sum_{i \in S(\epsilon)} \alpha_i e_i\| \leq \epsilon$$

with $\alpha_i = \langle f, e_i \rangle$ since the e_i's are orthonormal, and the approximation is measured in some norm, such as $\mathbb{L}^2$ or $\mathcal{H}^2$. The goal is to obtain representations in which the index set $S(\epsilon)$ in the summation is as small as possible, for a given approximation error ϵ. This hope is well founded at least when f is smooth or piecewise smooth, since in this case it should be compressible in some well-chosen basis $\{e_i\}$.

---

[1] There are several ways to define the inner product on the space of functions on a graph, including a weighted inner product that takes into account the invariant (stationary) distribution of the Markov chain induced by a random walk on the graph.

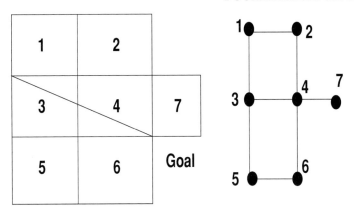

$$L = \begin{bmatrix} 2 & -1 & -1 & 0 & 0 & 0 & 0 \\ -1 & 2 & 0 & -1 & 0 & 0 & 0 \\ -1 & 0 & 3 & -1 & -1 & 0 & 0 \\ 0 & -1 & -1 & 4 & 0 & -1 & -1 \\ 0 & 0 & -1 & 0 & 2 & -1 & 0 \\ 0 & 0 & 0 & 0 & 0 & -1 & 1 \end{bmatrix}.$$

FIGURE 3.1: Top: A simple discrete state space can be modeled by an undirected unweighted graph, where each state is connected to its immediate neighbors. Bottom: Combinatorial graph Laplacian $L = D - W$, where W is the 0, 1 unweighted adjacency matrix, and D is a diagonal *valency* matrix of vertex degrees.

3.1.1 Function Approximation Using Laplacian Eigenfunctions

The *combinatorial Laplacian* L [26] is defined as $L = D - W$, where D is a diagonal matrix whose entries are the row sums of W. An example of the combinatorial Laplacian is given in Figure 3.1. Often, one considers the *normalized* Laplacian $\mathcal{L} = D^{-\frac{1}{2}}(D - W)D^{-\frac{1}{2}}$, whose eigenvalues lie in $[0, 2]$ [26]. The Laplacian is an *operator* on the space of functions $\mathcal{F} : V \to \mathbb{R}$ on a graph. In particular, it can be easily shown that

$$Lf(u) = \sum_{u \sim v}(f(u) - f(v))w(u, v) ,$$

that is, the Laplacian acts as a *difference* operator. On a two-dimensional grid, the Laplacian can be shown to essentially be a discretization of the continuous Laplace operator

$$\frac{\partial^2 f}{\partial x^2} + \frac{\partial^2 f}{\partial y^2},$$

where the partial derivatives are replaced by finite differences. Another fundamental property of the graph Laplacian is that projections of functions on the eigenspace of the Laplacian produce the smoothest global approximation respecting the underlying graph topology:

$$\langle f, Lf \rangle = \sum_{u \sim v} w_{uv}(f(u) - f(v))^2, \tag{3.2}$$

where this so-called *Dirichlet sum* is over the (undirected) edges $u \sim v$ of the graph G, and w_{uv} denotes the weight on the edge. Note that each edge is counted only once in the sum. From the standpoint of regularization, this property is crucial since it implies that rather than smoothing using properties of the ambient Euclidean space, smoothing takes the underlying manifold (graph) into account. This Laplacian is related to the notion of smoothness as above, since as we will show below:

$$\langle f, Lf \rangle = \sum_{u} f(u) \, Lf(u) = \sum_{u,v} w(u, v)(f(u) - f(v))^2 = ||\nabla f||_2^2 \,,$$

which should be compared with (3.1). Thus, one of the key attractive properties of the (combinatorial or normalized) Laplacian is that it is positive semi-definite. Since both the Laplacian operators, $\mathcal{L}$ and L, are also symmetric, the spectral theorem from Chapter 2 can be applied, yielding a discrete set of eigenvalues that are all non-negative: $0 \leq \lambda_0 \leq \lambda_1 \leq \cdots \lambda_i \leq \cdots$ and a corresponding orthonormal basis of real-valued eigenfunctions $\{\xi_i\}_{i \geq 0}$, solutions to the eigenvalue problem $\mathcal{L}\xi_i = \lambda_i \xi_i$.

The eigenfunctions of the Laplacian can be viewed as an orthonormal basis of global smooth functions that can be used for approximating any function on a graph [26]. A striking property of these basis functions is that they capture large-scale features of a graph, and are particularly sensitive to "bottlenecks", a phenomenon widely studied in the Riemannian geometry and spectral graph theory [23, 44, 26].

Observe that ξ_i satisfies $||\nabla \xi_i||_2^2 = \lambda_i$. In fact, the variational characterization of eigenvectors (described below) shows that ξ_i is the normalized function orthogonal to $\xi_0, \ldots, \xi_{i-1}$ with minimal $||\nabla \xi_i||_2$. Hence the projection of a function f on S onto the top k eigenvectors of the Laplacian is the smoothest approximation to f, in the sense of the norm in $\mathcal{H}^2$. A potential drawback of Laplacian approximation is that it detects only global smoothness, and may poorly approximate a function which is not globally smooth but only piecewise smooth, or with different smoothness in different regions. These drawbacks are addressed in the context of analysis with diffusion wavelets described in Chapter 4, and in fact partly motivated their construction [30].

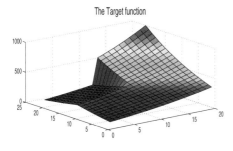

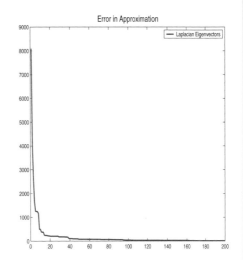

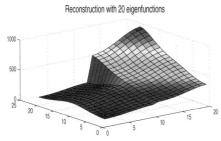

FIGURE 3.2: Target function (upper left), a vector $\in \mathbb{R}^{420}$, and its reconstruction using 20 eigenvectors of the normalized Laplacian (lower left). The error in reconstruction is plotted on the right.

3.1.2   Analysis–Synthesis Example

Figure 3.2 presents an example of function approximation using the eigenfunctions of the normalized Laplacian. This example comes from state-space planning (described in more detail in Chapter 6), where the solution to an optimization problem requires constructing a function called a *value function*. As shown in the figure, Laplacian eigenfunctions are able to approximate the desired function very efficiently.

3.2   RANDOM WALKS AND THE LAPLACIAN

Harmonic analysis on graphs and other discrete spaces can be carried out by diagonalization or dilation of a diffusion model [28, 29]. A diffusion model is intended to capture information flow on a graph or a *manifold*.[2] A simple diffusion model is a random walk on an undirected graph, where the probability of transitioning from a vertex to its neighbor is proportional to its degree, that is $P_r = D^{-1}W$. The Laplacian operators L and $\mathcal{L}$ defined in the previous section are closely related spectrally to the random walk operator P_r. The random walk matrix P_r is called a *diffusion model* because given any function f on the underlying graph G, the powers of $P_r^t f$ determine how quickly the random walk will "mix" and converge to the long term distribution. It can be shown that a random walk on an undirected graph defines a reversible

---

[2]Manifolds are defined more formally in Chapter 5.

Markov chain whose stationary distribution at a given vertex is given by $P(v) = \frac{d_v}{\text{vol}(G)}$, where d_v is the degree of vertex v and the "volume" $\text{vol}(G) = \sum_{v \in G} d_v$. Since the random walk matrix P_r is not symmetric, it is convenient to find a symmetrized diffusion model closely related to it spectrally. This is essentially the graph Laplacian matrix, which we introduced above.

To see the connection between the normalized Laplacian and the random walk matrix $P_r = D^{-1}W$, note the following identities:

$$\mathcal{L} = D^{-\frac{1}{2}} L D^{-\frac{1}{2}} = I - D^{-\frac{1}{2}} W D^{-\frac{1}{2}} \tag{3.3}$$
$$I - \mathcal{L} = D^{-\frac{1}{2}} W D^{-\frac{1}{2}} \tag{3.4}$$
$$D^{-\frac{1}{2}}(I - \mathcal{L})D^{\frac{1}{2}} = D^{-1}W. \tag{3.5}$$

Hence, the random walk operator $D^{-1}W$ is similar to $I - \mathcal{L}$, so both have the same eigenvalues, and the eigenvectors of the random walk operator are the eigenvectors of $I - \mathcal{L}$ point-wise multiplied by $D^{-\frac{1}{2}}$. The normalized Laplacian $\mathcal{L}$ also acts as a *difference* operator on a function f on a graph, that is

$$\mathcal{L}f(u) = \frac{1}{\sqrt{d_u}} \sum_{v \sim u} \left(\frac{f(u)}{\sqrt{d_u}} - \frac{f(v)}{\sqrt{d_v}} \right) w_{uv}. \tag{3.6}$$

The difference between the combinatorial and normalized Laplacian is that the latter models the degree of a vertex as a local measure.

3.2.1 Variational Analysis of Laplacian Eigenfunctions

Building on the Dirichlet sum above, a standard *variational* characterization of eigenvalues and eigenvectors views them as the solution to a sequence of minimization problems. In particular, the set of eigenvalues can be defined as the solution to a series of minimization problems using the *Rayleigh quotient* [26]. This provides a variational characterization of eigenvalues using projections of an arbitrary function $g : V \to \mathcal{R}$ onto the subspace $\mathcal{L}g$. The quotient gives the eigenvalues and the functions satisfying orthonormality are the eigenfunctions:

$$\frac{\langle g, \mathcal{L}g \rangle}{\langle g, g \rangle} = \frac{\langle g, D^{-\frac{1}{2}} L D^{-\frac{1}{2}} g \rangle}{\langle g, g \rangle} = \frac{\sum_{u \sim v}(f(u) - f(v))^2 w_{uv}}{\sum_u f^2(u) d_u},$$

where $f \equiv D^{-\frac{1}{2}} g$.

The first eigenvalue is $\lambda_0 = 0$, and is associated with the constant function $f(u) = \mathbf{1}$, which means the first eigenfunction $g_o(u) = \sqrt{D}\,\mathbf{1}$ (for an example of this eigenfunction, see the top-left plot in Figure 3.3). The first eigenfunction (associated with an eigenvalue 0) of the

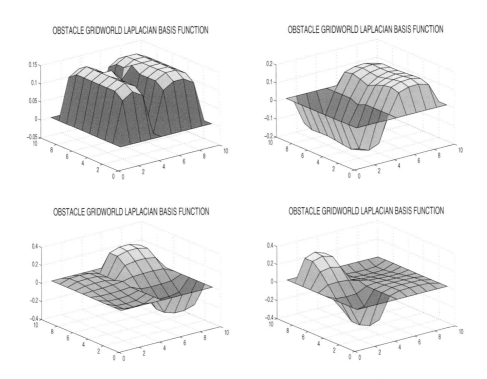

FIGURE 3.3: First four eigenfunctions of the normalized Laplacian for a "two-room" spatial environment modeled by an undirected graph with 100 vertices, divided into 57 accessible vertices (including one doorway vertex), and 43 inaccessible states representing exterior and interior walls (which are "one-vertex" thick).

combinatorial Laplacian is the constant function **1**. The second eigenfunction is the infimum over all functions $g : V \rightarrow \mathcal{R}$ that are perpendicular to $g_o(u)$, which gives us a formula to compute the first nonzero eigenvalue λ_1, namely

$$\lambda_1 = \inf_{f \perp \sqrt{D}\mathbf{1}} \frac{\sum_{u \sim v}(f(u) - f(v))^2 w_{uv}}{\sum_u f^2(u)d_u}.$$

The Rayleigh quotient for higher-order basis functions is similar: each function is perpendicular to the subspace spanned by previous functions (see the four plots in Figure 3.3). In other words, the eigenvectors of the graph Laplacian provide a systematic organization of the space of functions on a graph that respects its topology.

3.3   DIRECTED GRAPH LAPLACIAN

In this section we give a brief summary of the Laplacian on directed graphs [27]. A weighted directed graph is defined as $G_d = (V, E_d, W)$. The major distinction between the directed and undirected graphs is the non-reversibility of the edges. A directed graph may have weights $w_{ij} = 0$ and $w_{ji} \neq 0$. This is not possible in an undirected graph.

In order to define the graph Laplacians on G_d we must first introduce the Perron vector, ψ. The directed random walk transition matrix of G_d is defined as $P_d = D^{-1}W$. The Perron–Fröbenius theorem states that if G_d is strongly connected then P_d has a unique left eigenvector ψ with all positive entries such that $\psi P_d = \rho \psi$, where ρ is the spectral radius. ρ can be set to 1 by normalizing ψ such that $\sum_i \psi_i = 1$. A more intuitive way of thinking of ψ is as the long-term steady-state probability of being in any vertex at the end of a random walk on the graph.

There is no closed-form solution for ψ, however, there are several algorithms to calculate it. The *power method* is an approach to iteratively calculate ψ that starts with an initial guess for ψ, uses the definition $\psi P_d = \psi$ to determine a new estimate and iterates. Another technique is the Grassman–Taksar–Heyman (GTH) algorithm [48]. This technique uses a Gaussian elimination procedure designed to be numerically stable. The naive GTH implementation runs in $O(n^3)$, but this can be improved in $O(nm^2)$ if P_d is sparse. Other techniques, such as Perron complementation, have been introduced to speed up convergence [87].

The graph Laplacians for a directed graph are defined as [27]

$$L_d = \Psi - \frac{\Psi P_d + P_d^T \Psi}{2},$$

$$\mathcal{L}_d = I - \frac{\Psi^{1/2} P_d \Psi^{-1/2} + \Psi^{-1/2} P_d^T \Psi^{1/2}}{2},$$

where Ψ is a diagonal matrix whose entries are the Perron vector components $\Psi_{ii} = \psi_i$.

3.4   GRAPH PARTITIONING AND CHEEGER CONSTANTS

Many applications of graph theory use the properties of the *Fiedler* eigenvector [44] (the eigenvector associated with the smallest nonzero eigenvalue of the combinatorial or normalized Laplacian), such as its sensitivity to bottlenecks, in order to find clusters in graphs or partition them. To formally explain this, we briefly review spectral geometry. The *Cheeger* constant h_G of a graph G is defined as [24, 26]

$$h_G(S) = \min_S \frac{|E(S, \tilde{S})|}{\min(\text{vol } S, \text{vol } \tilde{S})}.$$

Here, S is a subset of vertices, $\tilde{S}$ is the complement of S, and $E(S, \tilde{S})$ denotes the set of all edges (u, v) such that $u \in S$ and $v \in \tilde{S}$. The volume of a subset S is defined as vol $S = \sum_{x \in S} d_x$. Consider the problem of finding a subset S of states such that the edge boundary ∂S contains as few edges as possible, where

$$\partial S = \{(u, v) \in E(G) : u \in S \text{ and } v \notin S\}.$$

The relation between ∂S and the Cheeger constant is given by

$$|\partial S| \geq h_G \text{ vol } S.$$

In the two-room spatial environment illustrated in Figure 3.3, the Cheeger constant is minimized by setting S to be the vertices corresponding to locations in the first room, since this will minimize the numerator $E(S, \tilde{S})$ and maximize the denominator min(vol S, vol $\tilde{S}$). A remarkable identity connects the Cheeger constant with the spectrum of the graph Laplacian operator. This theorem underlies the reason why the eigenfunctions associated with the second eigenvalue λ_1 of the graph Laplacian captures the geometric structure of environments, as illustrated in Figure 3.3.

Theorem 3.1. *[26] Define λ_1 to be the first (nonzero) eigenvalue of the normalized graph Laplacian operator $\mathcal{L}$ on a graph G. Let h_G denote the Cheeger constant of G. Then, we have $2h_G \geq \lambda_1 > \frac{h_G^2}{2}$.*

3.5 BIBLIOGRAPHICAL REMARKS

A comprehensive treatment of the spectra and eigenspaces of graphs based on the adjacency matrix is given in [31, 32]. Chung [26] gives a detailed introduction to spectral graph theory, focusing extensively on the spectrum of the normalized Laplacian. An up-to-date overview of the properties of the eigenvectors of the graph Laplacian is given in [119]. The directed Laplacian is defined in [27].

CHAPTER 4

Multiscale Bases on Graphs

For all its successes, Fourier analysis has some significant limitations. The basis functions—for example, the eigenvectors of a graph Laplacian—are localized in frequency, but their support is *global*. Consequently, Fourier bases are relatively poor at approximating piecewise-smooth functions with local discontinuities. Fourier analysis also does not reveal multiscale regularities. These limitations of Fourier analysis have only recently been overcome through the collaborative effort of engineers, mathematicians, and scientists working over the past two decades. The resulting framework, popularly called *wavelets* [34, 82], can be described akin to the design of a powerful new mathematical microscope, probing and revealing the properties of functions and sets at multiple temporal and spatial scales.

In this chapter, we describe a specific approach to wavelets analysis in discrete spaces, such as graphs, called *diffusion wavelets* [30, 21]. This approach forms a parallel in many respects with the extension of Fourier analysis to graphs described in Chapter 3, where the graph Laplacian was diagonalized to find basis functions. Here, the approach to constructing bases follows the wavelet tradition of *dilation* [82], where a basis function is dilated in time. What does it mean to dilate a function on a graph? The answer proposed in [30] is to define dilations through the application of a diffusion operator, such as a random walk. The resulting multi-resolution analysis produces both scaling functions and wavelets at multiple temporal and spatial scales. These basis functions provide a really interesting way to compress powers of transition matrices, which is often an essential computational step in many application domains.

4.1 INTRODUCTION

This chapter introduces a novel multiscale framework called *diffusion wavelets* [30, 21]. Diffusion wavelet bases are adapted to the geometry of a graph, and can be learned adaptively from sampling a data set or a state space. The wavelet framework provides significant advantages in that the constructed bases have compact support, and the approach yields a novel approach to the hierarchical abstraction of stochastic processes such as Markov chains and Markov decision processes [78, 72, 98]. In Chapter 3, Fourier analysis was associated with expansions on eigenfunctions of the Laplacian. This framework is a powerful tool for the *global* analysis of functions, however it is known to be relatively poor at modeling or approximating local or transient properties [81]. This motivated the construction, about 20 years ago, of classical wavelets, which allow a very efficient *multiscale* analysis, much like a powerful tunable microscope probing the properties of a function at different locations and scales. Recently wavelet analysis has been generalized in a natural way to manifolds and graphs, and these techniques, termed *diffusion wavelets* because they are associated with a diffusion process [28, 29] that defines the different scales, allow a multiscale analysis of functions on manifolds and graphs.

Diffusion wavelets have desirable properties in view of applications to learning, function approximation, compression and denoising of functions on graphs and manifolds. In many applications the multiscale diffusion wavelet analysis constructed is interpretable and meaningful. In Chapter 6, we use diffusion wavelets to construct basis functions for solving Markov decision processes, where it results in new aggregate groupings of states and actions. In Chapter 7, we use diffusion wavelets to construct new multiscale methods for the compression of 3D objects in computer graphics, where the basis functions capture meaningful semantic regions of objects. Finally, in Chapter 8, we apply this framework to the analysis of document corpora, where it yields groupings of documents (or words) at different scales, corresponding to topics at different levels of specificity.

Diffusion wavelets enable a hierarchical analysis of functions on a graph by constructing a multiscale tree of *wavelet*-type basis functions on the graph, which allows efficient hierarchical representation of not just functions, but also yields a fast algorithm for the inversion of operators like the random walk or the Laplacian.

4.2   MULTI-RESOLUTION ANALYSIS AND SYNTHESIS

Wavelets generalize the analysis–synthesis perspective introduced previously to the case where both the analysis—the "measurements" of a function f by linear functionals $\langle f, \phi_i \rangle$—and the reconstruction or synthesis phase are carried out at multiple spatial or temporal scales.

Consider a one-dimensional function f (e.g. a signal, such as a sound). We want to efficiently represent such a function, or perform tasks such as compression or denoising. *Transform methods* use a (usually linear) invertible map $f \mapsto \hat{f}$, where this map ideally has the property that simple operations on $\hat{f}$, followed by an inversion of the transformation, can be used to perform the task at hand. We illustrated this process in Chapter 3, where the eigenfunctions of the graph Laplacian were used to analyze and then reconstruct a function on a graph. Generally speaking, let $\hat{f}$ be the set of coefficients of f onto an orthonormal basis (e.g. Fourier or Laplacian), and simple operations include hard- and soft-thresholding (e.g., setting to 0 all the coefficients below a certain threshold τ). When the function f is expected to have different behavior at different locations (for example a value function in a Markov decision process [98] or a coordinate function for a 3D object [59]), it is natural to analyze and transform such a function using basis functions which are localized. Since the amount of localization is unknown, and may change from location to location, it is desirable to have basis functions localized at all possible scales: the ones at coarse scale analyze slow variations in the signal, while the ones at fine scale analyze more rapid variations. Wavelets are an example of such a basis.

A consolidated framework in wavelet analysis is the idea of *multi-resolution analysis* (MRA) [34, 82]. A multi-resolution analysis of the space of square-integrable functions $\mathbb{L}^2(\mathbb{R})$ is a sequence of subspaces $\{V_j\}_{j\in\mathbb{Z}}$ with the following properties:

(i) $V_{j+1} \subseteq V_j, \overline{\cup_{j\in\mathbb{Z}} V_j} = \mathbb{L}^2(\mathbb{R}), \cap_{j\in\mathbb{Z}} V_j = \{0\}$;
(ii) $f \in V_{j+1}$ if and only if $f(2\cdot) \in V_j$;
(iii) there exists an orthonormal basis $\{\varphi_{j,k}\}_{k\in\mathbb{Z}} := \{2^{-\frac{j}{2}}\varphi(2^{-j}\cdot -k)\}_{k\in\mathbb{Z}}$ of V_j.

The subspace V_j is called the jth approximation or *scaling* space. The functions $\varphi_{j,k}$ are called *scaling functions*. The function φ that generates, under dyadic (power of two) dilations and integer translations, the family $\varphi_{j,k}$ is called the mother scaling function. The *multi-scale orthogonal projection* on the scaling space V_j

$$P_j f = \sum_{k\in\mathbb{Z}} \langle f, \varphi_{j,k}\rangle \varphi_{j,k} \tag{4.1}$$

gives an approximation at scale j. As j increases these approximations get coarser and coarser, while as j decreases the approximations get finer and finer, and eventually (because of (ii)), they tend to f: $P_j f \to f$ as $j \to -\infty$, where the limit is taken in $\mathbb{L}^2(\mathbb{R})$. One says that $P_j f$ is an approximation of f at scale j.

4.2.1   Multi-resolution Analysis on Graphs

We now specialize the above general multi-resolution analysis framework to functions on graphs. Let T be an operator on a graph, such as the random walk $T = D^{-1}W$. The key idea underlying multiscale analysis on graphs is to generate basis functions across multiple scales using T as a *dilation* operator. Define dyadic spatial scales t_j as

$$t_j = \sum_{t=0}^{j} 2^t = 2^{j+1} - 1, \quad j \geq 0 .$$

We can define "low-pass" subspaces by selecting those eigenvalues in the spectrum of T (which lies in $[0, 1]$) that are above a given threshold, where for higher levels, we take the corresponding powers. More precisely, define the spectral subspace

$$\sigma_j(T) = \{\lambda \in \sigma(T), \lambda^{t_j} \geq \varepsilon\}$$

where $\varepsilon \in (0, 1)$ is a pre-defined threshold. We can now associate with each of the spectral subspaces a vector subspace of eigenvectors associated with each spectral subspace:

$$V_j = \langle\{\xi_\lambda : \lambda \in \sigma(T), \lambda^{t_j} \geq \varepsilon\}\rangle, \quad j \geq 0 .$$

By definition, we let $V_{-1} = \mathbb{L}^2(G)$, the set of all functions on the graph G. In the limit, we obtain

$$\lim_{j \to \infty} V_j = \langle\{\xi_\lambda : \lambda_i = 1\}\rangle,$$

that is, the space spanned by the eigenvectors associated with the largest eigenvalue of T. We now illustrate this framework with two examples, showing the basis functions spanning the V_j subspaces, and then proceed to showing how they can be efficiently computed.

4.2.2   Examples of Diffusion Wavelets

Before we give a mathematically rigorous derivation of the diffusion wavelet framework, it will help us to consider some examples. In Chapter 3, we considered example graphs representing a spatial environment of two rooms with a connecting door serving as a "bottleneck" between the rooms. Figure 4.1 shows examples of scaling function bases constructed by the diffusion wavelet algorithm. The figure shows scaling functions from a ten-level diffusion wavelet tree. At the lowest level (the top-left plot in the figure), the scaling functions are just the "delta" unit vector bases. At subsequent levels, the unit vector bases are "dilated" using the random walk on the graph, and the resulting dilated vectors are then orthogonalized to construct the scaling functions. Note how the scaling functions get "coarser" at each succeeding level, till at the topmost level (the bottom-right plot), they look like the eigenvectors shown in Chapter 3.

We consider another example in Figure 4.2. This graph represents a spatial environment, in this case a single room with a square "obstacle" placed in the center representing a set of unreachable vertices. Once again, at the bottom of the hierarchy, the scaling functions are just the "delta" unit vector bases (the top-left plot), which are progressively dilated using powers of the random walk. Note how the structure of the "obstacle" is clearly revealed at the higher levels.

4.3   DIFFUSION ANALYSIS

For the purpose of this discussion, we restrict our attention to the case of a finite undirected weighted graph (G, E, W). If P represents one step of the random walk $D^{-1}W$, by the Markov property P^t represents t steps. For an initial condition δ_x (i.e, where x is the starting state), $P^t\delta_x(y)$ represents the probability of being at y at time t, conditioned on starting in state x. The matrix P encodes local similarities between points, and the matrix P^t is diffusing, or

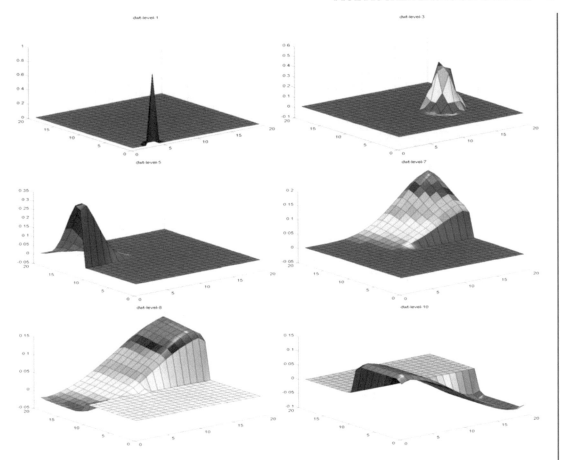

FIGURE 4.1: Diffusion wavelet scaling functions on a graph representing a spatial "two-room" environment.

integrating, this local information for t steps to larger and larger neighborhoods of each point. The process $\{P^t\}_{t\geq 0}$ can be analyzed at different time scales. For very large times, the random walk can be analyzed through its top eigenvectors, which are related to those of a Laplacian on the graph/manifold. The analysis for large times leads to *Fourier analysis* of the large-scale spatial and temporal regularities of the graph/manifold, and to the identification of useful structures, such as large-scale clusters.

For small and medium times, the random walk in general cannot be studied effectively with Laplacian eigenfunctions, which are global and not suited for analyzing the small- and medium-scale behavior. On the other hand, many interesting features of the data and of functions on the data can be expected to exist at small and medium time scales: one remarkable example is complex (computer, biological, information, social) networks, where communities

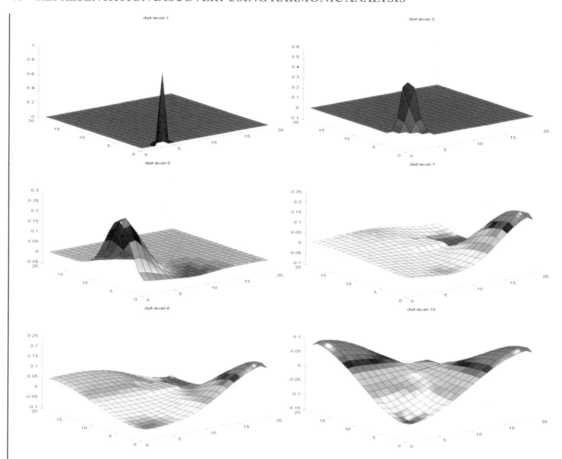

FIGURE 4.2: Diffusion wavelet scaling functions on a graph representing a spatial "room" environment with a central "obstacle" region.

(of computers, genes and/or proteins, people) of different sizes co-exist and cluster together at different scales. Another important example is the boundary between two classes in a classification problem. It is easy to imagine regions of the state space where the process is smoother than others. The task of analyzing P^t for all times and locations seems tantalizing, since it would seem to require either large time in order to compute all powers of P (which is computationally expensive since, even if P is sparse, its powers are not), and/or large space to store those powers. However, it is easy to observe that there is redundancy in time and space in the family $\{P^t(x, y)\}_{t \geq 0; x, y \in X}$. First of all there is a spatial redundancy: if x and y are close and t is large (depending on the distance between x and y), $P^t(x, \cdot)$ is very similar to $P^t(y, \cdot)$ (as distributions on X). Second, there is a redundancy across time scales: if we know $P^t(x, \cdot)$ and $P^t(y, \cdot)$, then by the Markov property we know $P^{2t}(x, y)$. It is remarkable that this redundancy can be eliminated, and an efficient multiscale encoding is possible. This leads to *diffusion wavelet analysis* [30].

4.3.1 Basic Setup and Notation

We reintroduce notation from the previous chapters that will be useful in this chapter. $x \sim y$ means that there is an edge between vertices x and y, $d(x) = \sum_{x \sim y} w(x, y)$ is the degree of x, D is the diagonal matrix defined by $D_{xx} = d(x)$, and W is the matrix defined by $W_{xy} = w(x, y) = w(y, x)$. We can assume $w(x, y) > 0$ if $x \sim y$. Sometimes S is naturally endowed with a measure (weight) μ on its vertices. A typical example is $\mu(\{x\}) = d(x)$; in some other cases μ could be a probability distribution, for example related to sampling. In most of what follows we shall assume that μ is simply the counting measure, but the construction generalizes to the case of general measures μ. One defines the space of square-integrable functions

$$\mathbb{L}^2(G) := \{f : G \to \mathbb{R} \text{ s.t. } ||f||_2^2 := \sum_{x \in G} |f(x)|^2 \mu(\{x\}) < +\infty\},$$

which is a Hilbert space with the natural inner product (associated with $|| \cdot ||_2$)

$$\langle f, g \rangle = \sum_{x \in G} f(x)g(x)\mu(\{x\}).$$

There is a natural random walk on G, given by $P = D^{-1}W$. This Markov chain is necessarily reversible, and thus it is conjugate, together with its powers, to a symmetric operator T:

$$\begin{aligned}
T^t &= D^{\frac{1}{2}} P^t D^{-\frac{1}{2}} = (D^{-\frac{1}{2}} W D^{-\frac{1}{2}})^t \\
&= (I - \mathcal{L})^t = \sum_i (1 - \lambda_i)^t \xi_i(\cdot)\xi_i(\cdot),
\end{aligned} \tag{4.2}$$

where

$$\mathcal{L} = D^{-\frac{1}{2}}(D - W)D^{-\frac{1}{2}} \tag{4.3}$$

is the normalized Laplacian, and $0 = \lambda_0 \leq \lambda_1 \leq \cdots \leq \lambda_i \leq \cdots$ are the eigenvalues of $\mathcal{L}$ and $\{\xi_i\}$ the corresponding eigenvectors: $\mathcal{L}\xi_i = \lambda_i \xi_i$. Clearly $P^t = D^{-\frac{1}{2}} T^t D^{\frac{1}{2}}$, and hence studying T is equivalent, as far as spectral properties are concerned, to studying P.

4.3.2 Multiscale Analysis of Functions and Stochastic Processes

In this section, we view multiscale analysis from two related, but nonetheless distinct perspectives. The first is an approximation of functions, the other is an approximation of stochastic (Markov) processes. The multiscale and wavelet analysis of functions is well understood in

Euclidean space, and is motivated by the need for studying functions (or signals) that have different behavior at different locations at different scales.

Regarding multiscale analysis of stochastic processes, many applications require representing time series data at multiple levels of resolution, for example robot navigation [117], sensor networks [49], and social network analysis [85]. Given the inherent uncertainty in such domains, a computational approach that automatically abstracts stochastic processes at multiple levels of abstraction is highly desirable. The diffusion wavelet framework provides a general and powerful way of learning multiscale structures, relieving a human of having to hand code a suitable hierarchical structure. In particular, diffusion wavelets and other multiscale techniques on graphs enable automatically constructing basis representations at multiple levels of abstraction of a diffusion-like process.

4.4 DIFFUSION WAVELETS

Diffusion wavelets enable a fast multiscale analysis of functions on a manifold or graph, generalizing wavelet analysis and associated signal processing techniques (such as compression or denoising) to functions on manifolds and graphs. They allow the efficient and accurate computation of high powers of a Markov chain P on the manifold or graph, including the direct computation of the Green's function (or fundamental matrix) of the Markov chain, $(I - P)^{-1}$.

A multi-resolution decomposition of the functions on the graph is a family of nested subspaces $V_0 \supseteq V_1 \supseteq \cdots \supseteq V_j \supseteq \cdots$ spanned by orthogonal bases of diffusion scaling functions Φ_j. If we interpret T^t as an operator on functions on the graph, then V_j is defined as the numerical range, up to the precision ε, of $T^{2^{j+1}-1}$, and the scaling functions are smooth bump functions with some oscillations, at scale roughly 2^{j+1} (measured with respect to the geodesic distance). The orthogonal complement of V_{j+1} into V_j is called W_j, and is spanned by a family of orthogonal diffusion wavelets Ψ_j, which are smooth localized oscillatory functions at the same scale.

4.4.1 Construction of Diffusion Wavelets

Here and in the rest of this section, we will reuse the notation introduced in Chapter 2, where $[L]_{B_1}^{B_2}$ indicates the matrix representing the linear operator L with respect to the basis B_1 in the domain and B_2 in the range. A set of vectors B_1 represented on a basis B_2 will be written in the matrix form $[B_1]_{B_2}$, where the rows of $[B_1]_{B_2}$ are the coordinates of the vectors B_1 in the coordinate system defined by B_2.

The input to the algorithm is a "precision" parameter $\varepsilon > 0$, and a weighted graph (G, E, W). We assume that G is strongly connected and local, in the sense that each vertex is connected to a small number of vertices. The construction is based on using the natural random

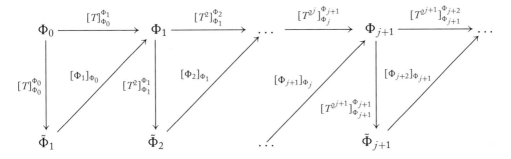

FIGURE 4.3: The figure shows downsampling, orthogonalization, and operator compression. (All triangles are commutative by construction.)

walk $P = D^{-1}W$ on a graph (where D is the out-degree matrix if the graph is directed) which we assume aperiodic. We use the powers of P to "dilate", or "diffuse" functions on the graph, and then define an associated coarse-graining of the graph. Observe that in many cases of interest P is a sparse matrix. We usually *normalize* P and consider $T = \Pi P \Pi^{-1}$, where Π is the asymptotic distribution of P, which by the hypotheses on P exists, is unique and can be chosen to be a strictly positive distribution by the Perron–Fröbenius theorem. If G is undirected, P is reversible, $\Pi = D^{\frac{1}{2}}$, and T is symmetric. In the other cases, if T is not symmetric, in what follows any statement regarding eigenvectors should be disregarded.

We assume that T is a sparse matrix, and that the numerical rank of the powers of T decays rapidly with the power. For example, a desirable situation is when the number of singular values of T^t larger than ε is smaller than $2^{-\gamma t}$. A diffusion wavelet tree consists of orthogonal diffusion scaling functions Φ_j that are smooth bump functions, with some oscillations, at scale roughly 2^j (measured with respect to the geodesic distance), and orthogonal wavelets Ψ_j that are smooth localized oscillatory functions at the same scale. The scaling functions Φ_j span a subspace V_j, with the property that $V_{j+1} \subseteq V_j$, and the span of Ψ_j, W_j, is the orthogonal complement of V_j into V_{j+1}. This is achieved by using the dyadic powers T^{2^j} as "dilations", to create smoother and wider (always in a geodesic sense) "bump" functions (which represent densities for the symmetrized random walk after 2^j steps), and orthogonalizing and downsampling appropriately to transform sets of "bumps" into orthonormal scaling functions.

We now describe the multiscale construction briefly, and further details can be found in the original paper [30]. It may be useful to compare the description that follows with the diagram in Figure 4.3. T is initially represented on the basis $\Phi_0 = \{\delta_k\}_{k \in G}$; we consider the columns of T, interpreted as the set of functions $\tilde{\Phi}_1 = \{T\delta_k\}_{k \in G}$ on G. A local multiscale orthogonalization procedure is used to carefully orthonormalize these columns to get a basis $\Phi_1 = \{\varphi_{1,k}\}_{k \in G_1}$ (G_1 is *defined* as this index set), written with respect to the basis Φ_0, for the range of T up to the precision ε. This information is stored in the sparse matrix $[\Phi_1]_{\Phi_0}$. This yields a subspace that we denote by V_1. Essentially, Φ_1 is a basis for the subspace V_1 which is ε-close to the range of T, and with basis elements that are well localized. Moreover, the elements of Φ_1 are coarser than the elements of Φ_0, since they are the result of applying the "dilation" T once. Obviously $|G_1| \leq |G|$, but this inequality may already be strict since the numerical range of T may be approximated, within the specified precision ε, by a subspace of smaller dimension. Whether this is the case or not, we have computed the sparse matrix $[T]_{\Phi_0}^{\Phi_1}$, a representation of an ε-approximation of T with respect to Φ_0 in the domain and Φ_1 in the range. We can also represent T in the basis Φ_1: with the notation above this is the matrix $[T]_{\Phi_1}^{\Phi_1}$. We compute $[T^2]_{\Phi_1}^{\Phi_1} = [\Phi_1]_{\Phi_0}[T^2]_{\Phi_0}^{\Phi_0}[\Phi_1]_{\Phi_0}^T$. If T is self-adjoint, this is equal to $[T]_{\Phi_0}^{\Phi_1}([T]_{\Phi_0}^{\Phi_1})^T$, which has the advantage that numerical symmetry is forced upon $[T^2]_{\Phi_1}^{\Phi_1}$. In the general (non-symmetric) case, $[T^2]_{\Phi_1}^{\Phi_1} = ([T]_{\Phi_0}^{\Phi_1}[\Phi_1]_{\Phi_0})^2$.

It is now clear how to proceed: we look at the columns of $[T^2]_{\Phi_1}^{\Phi_1}$, which are $\tilde{\Phi}_2 = \{[T^2]_{\Phi_1}^{\Phi_1}\delta_k\}_{k \in G_1}$. By unraveling the notation, these are functions $\{T^2\varphi_{1,k}\}_{k \in G_1}$, up to the precision ε. Once again we apply a local orthonormalization procedure to this set of functions, obtaining an orthonormal basis $\Phi_2 = \{\varphi_{2,k}\}_{k \in G_2}$ for the range of T_1^2 (up to the precision ε), and also for the range of T_0^3 (up to the precision 2ε). Observe that Φ_2 is naturally written with respect to the basis Φ_1, and hence encoded in the matrix $[\Phi_2]_{\Phi_1}$. Moreover, depending on the decay of the spectrum of T, $|G_2|$ is in general a fraction of $|G_1|$. The matrix $[T^2]_{\Phi_1}^{\Phi_2}$ is then of size $|G_2| \times |G_1|$, and the matrix $[T^4]_{\Phi_2}^{\Phi_2} = [T^2]_{\Phi_1}^{\Phi_2}([T^2]_{\Phi_1}^{\Phi_2})^T$, a representation of T^4 acting on Φ_2, is of size $|G_2| \times |G_2|$.

After j iterations in this fashion, we will have a representation of T^{2^j} onto a basis $\Phi_j = \{\varphi_{j,k}\}_{k \in G_j}$, encoded in a matrix $T_j := [T^{2^j}]_{\Phi_j}^{\Phi_j}$. The orthonormal basis Φ_j is represented with respect to Φ_{j-1}, and encoded in the matrix $[\Phi_j]_{\Phi_{j-1}}$. We let $\tilde{\Phi}_j = T_j\Phi_j$. We can represent the next dyadic power of T on Φ_{j+1} on the range of T^{2^j}. Depending on the decay of the spectrum of T, we expect $|G_j| << |G|$, in fact in the ideal situation the spectrum of T decays fast enough so that there exists $\gamma < 1$ such that $|G_j| < \gamma|G_{j-1}| < \cdots < \gamma^j|G|$. This corresponds to downsampling the set of columns of dyadic powers of T, thought of as vectors in $\mathbb{L}^2(G)$. The hypothesis that the rank of powers of T decreases guarantees that we can down-sample and obtain coarser and coarser lattices in these spaces of columns.

While Φ_j is naturally identified with the set of Dirac δ-functions on G_j, we can extend these functions living on the "compressed" (or "downsampled") graph G_j to the whole initial graph G by writing

$$[\Phi_j]_{\Phi_0} = [\Phi_j]_{\Phi_{j-1}}[\Phi_{j-1}]_{\Phi_0} = \cdots = [\Phi_j]_{\Phi_{j-1}}[\Phi_{j-1}]_{\Phi_{j-2}} \cdots [\Phi_1]_{\Phi_0}[\Phi_0]_{\Phi_0}. \qquad (4.4)$$

Since every function in Φ_0 is defined on G, so is every function in Φ_j. Hence any function on the compressed space G_j can be extended naturally to the whole G. In particular, one can

$\{\Phi_j\}_{j=0}^{J}, \{\Psi_j\}_{j=0}^{J-1}, \{[T^{2^j}]_{\Phi_j}^{\Phi_j}\}_{j=1}^{J} \leftarrow \texttt{DiffusionWaveletTree}\,([T]_{\Phi_0}^{\Phi_0}, \Phi_0, J, \mathrm{SpQR}, \varepsilon)$

// **Input:**

// $[T]_{\Phi_0}^{\Phi_0}$: a diffusion operator, written on the orthonormal basis Φ_0

// Φ_0 : an orthonormal basis which ε-spans V_0

// J : number of levels

// SpQR : function to compute a sparse QR decomposition.

// ε: precision

// **Output:**

// The orthonormal bases of scaling functions Φ_j, wavelets Ψ_j, and

// compressed representation of T^{2^j} on Φ_j for j in the requested range.

for $j = 0$ **to** $J - 1$ **do**

$\quad [\Phi_{j+1}]_{\Phi_j}$, $[T^{2^j}]_{\Phi_j}^{\Phi_{j+1}} \leftarrow \texttt{SpQR}([T^{2^j}]_{\Phi_j}^{\Phi_j}, \varepsilon)$

$\quad T_{j+1} := [T^{2^{j+1}}]_{\Phi_{j+1}}^{\Phi_{j+1}} \leftarrow ([T^{2^j}]_{\Phi_{j+1}}^{\Phi_{j+1}}[\Phi_{j+1}]_{\Phi_j})^2$

$\quad [\Psi_j]_{\Phi_j} \leftarrow \texttt{SpQR}(I_{\langle\Phi_j\rangle} - [\Phi_{j+1}]_{\Phi_j}[\Phi_{j+1}]_{\Phi_j}^{T}, \varepsilon)$

end

$\{Q, R\} \longleftarrow \texttt{SpQR}(A, \varepsilon)$

$\quad$ // A_j: the jth column of A.

$\quad k = 0; stop = 0; Q = \{\}; B = A;$

$\quad$ **while** $(stop \neq 1)$

$\quad\quad i \longleftarrow \arg_j \max(\|A_j\|_2);$

$\quad\quad$ **if** $(\|A_i\| < \varepsilon)$ $\{stop = 1;\}$

$\quad\quad$ **else**

$\quad\quad\quad k = k + 1;$

$\quad\quad\quad e_k = A_i/\|A_i\|;$

$\quad\quad\quad Q = Q\bigcup e_k; A = A \setminus A_i;$

$\quad\quad\quad$ Orthogonalize remaining elements of A to e_k, obtaining $\tilde{A}$;

$\quad\quad\quad A \longleftarrow \tilde{A};$

$\quad\quad$ **end if**

$\quad$ **end while**

$\quad R = Q^T B;$

FIGURE 4.4: Pseudo-code for the construction of a Diffusion Wavelet Tree.

compute low-frequency eigenfunctions on G_j in compressed form, and then extend them to the whole G. The elements in Φ_j are at scale $T^{2^{j+1}-1}$, and are much coarser and "smoother", than the initial elements in Φ_0, which is how they can be represented in compressed form. The projection of a function onto the subspace spanned by Φ_j will be by definition an approximation to that function at that particular scale.

There is an associated fast scaling function transform: suppose we are given f on G and want to compute $\langle f, \varphi_{j,k} \rangle$ for all scales j and corresponding "translations" k. Being given f means we are given $(\langle f, \varphi_{0,k} \rangle)_{k \in G}$. Then we can compute $(\langle f, \varphi_{1,k} \rangle)_{k \in G_1} = [\Phi_1]_{\Phi_0}(\langle f, \varphi_{0,k} \rangle)_{k \in G}$, and so on for all scales. The sparser the matrices $[\Phi_j]_{\Phi_{j-1}}$ (and $[T]_{\Phi_j}^{\Phi_j}$), the faster this computation. This generalizes the classical scaling function transform. Wavelet bases for the spaces W_j can be built analogously by factorizing $I_{V_j} - Q_{j+1} Q_{j+1}^T$, which is the orthogonal projection on the complement of V_{j+1} into V_j. The spaces can be further split to obtain wavelet packets [21]. The wavelets can be considered as high-pass filters, in the sense that they capture the detail lost from going from V_j to V_{j+1}, and also in the sense that their expansion in terms of eigenfunctions of the Laplacian essentially only involves eigenfunctions corresponding to eigenvalues in $[\varepsilon^{-2^j-1}, \varepsilon^{-2^{j+1}-1}]$. In particular, their Sobolev norm, or smoothness, is controlled.

In the same way, any power of T can be applied efficiently to a function f. Also, the Green's function $(I - T)^{-1}$ can be applied efficiently to any function, since it can be represented as the product of the dyadic powers of T, each of which can be applied efficiently. We are at the same time compressing the powers of the operator T and the space itself, at essentially the optimal "rate" at each scale, as dictated by the portion of the spectrum of the powers of T which is above the precision ε.

Observe that each point in G_j can be considered as a "local aggregation" of points in G_{j-1}, which is completely dictated by the action of the operator T on functions on G: the operator itself is dictating the geometry with respect to which it should be analyzed, compressed, or applied to any vector. The algorithm is summarized in Figure 4.4.

A fast diffusion wavelet transform allows expanding in $\mathcal{O}(n)$ computations (where n is the number of vertices) any function in the wavelet, or wavelet packet, basis, and efficiently search for the most suitable basis set. Diffusion wavelets and wavelet packets are a very efficient tool for the representation and approximation of functions on manifolds and graphs [30, 21], generalizing to these general spaces the nice properties of wavelets that have been so successfully applied in Euclidean spaces.

Diffusion wavelets allow computing $T^{2^k} f$ for any fixed f, in order $\mathcal{O}(kn)$. This is non-trivial because while the matrix H is sparse, large powers of it are not, and the computation $T^{2^k} f = T \cdot T \cdots (T(Tf)) \cdots)$ involves 2^k matrix–vector products. As a notable consequence, this yields a fast algorithm for computing the Green's function, or fundamental matrix,

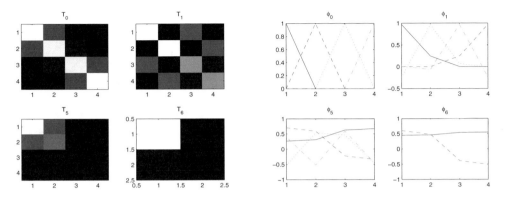

FIGURE 4.5: The four panels on the top display matrices $[T^{2^j}]_{\Phi_j}^{\Phi_j}$ representing compressed dyadic powers of T, with gray level representing entry values. Observe that the size of the matrix decays, since so does the rank of the powers of T. The four panels on the bottom illustrate some scaling function bases on the four-state Markov chain.

associated with the Markov process T, via the Schultz expansion [72]:

$$(I - T^1)^{-1}f = \sum_{k \geq 0} T^k = \prod_{k \geq 0}(I + T^{2^k})f.$$

In a similar way one can compute $(I - P)^{-1}$. For large classes of Markov chains we can perform this computation in time $\mathcal{O}(n)$, in a direct (as opposed to iterative) fashion. This is remarkable since in general the matrix $(I - T^1)^{-1}$ is full and just writing down the entries would take time $\mathcal{O}(n^2)$. It is the multiscale compression scheme that allows efficiently representing $(I - T)^{-1}$ in compressed form, taking advantage of the smoothness of the entries of the matrix.

4.4.2 Multiscale Compression of a Simple Markov Chain

To illustrate the multiscale analysis enabled by diffusion wavelets, it helps us to see the results of the analysis on a simple example. We consider the Markov chain on four states $\{a, b, c, d\}$:

$$T = \begin{pmatrix} 0.8 & 0.2 & 0 & 0 \\ 0.2 & 0.75 & 0.05 & 0 \\ 0 & 0.05 & 0.75 & 0.2 \\ 0 & 0 & 0.2 & 0.8 \end{pmatrix}.$$

This chain has a "bottleneck" between states $\{a, b\}$ and states $\{c, d\}$. We fix a precision $\varepsilon = 10^{-10}$. See Figure 4.5 for the discussion that follows. The scaling functions Φ_0 are simply $\{\delta_a, \delta_b, \delta_c, \delta_d\}$. We apply T to Φ_0 and orthonormalize to get Φ_1 (Figure 4.5). Each function in Φ_1 is an "abstract-state", i.e. a linear combination of the original states. We represent T^2 on Φ_1, to get a matrix T_2, apply to Φ_1 and orthonormalize, and so on. At scale 5, we have the

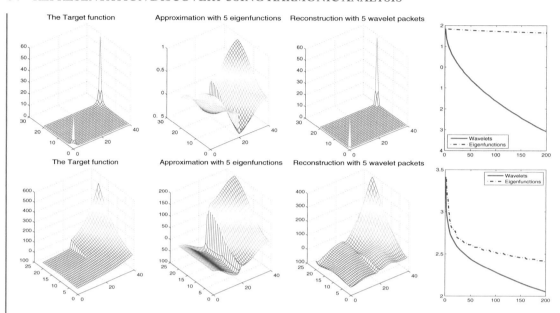

FIGURE 4.6: Left column: target functions. Middle two columns: approximations produced by five diffusion wavelet bases and Laplacian eigenfunctions. Right column: least-squares approximation error (the log scale) using up to 200 basis functions (bottom curve: diffusion wavelets; top curve: Laplacian eigenfunctions).

basis Φ_5 and the operator T_5, representing T^{2^5} on Φ_5. At the next level, we obtain Φ_7, which is only two dimensional, because $T_5 \Phi_5$ has ε-rank 2 instead of 4: of the four "abstract-states" $T_5 \Phi_5$, only two of them are at least ε-independent. Observe the two scaling functions in Φ_6 are approximately the asymptotic distribution and the function which distinguishes between the two clusters $\{a, b\}$ and $\{c, d\}$. Then T_6 represents T^{2^6} on Φ_7 and is a 2 by 2 matrix. At scale 10, Φ_{10} is one-dimensional, and is simply the top eigenvector of T (represented in compressed form, on the basis Φ_8), and the matrix T_9 is 1 by 1 and is just the top eigenvalue, 1, of T.

Already in this simple example we see that the multiscale analysis generates a sequence of Markov chains, each corresponding to a different time scale (i.e. power of the original Markov chain), represented on a set of scaling functions (aggregates of states) in compressed form.

4.4.3 Comparison of Eigenfunction and Diffusion Wavelet Bases

We end with an illustrative example showing where diffusion wavelet bases excel, and where Fourier bases of the Laplacian do really poorly. The top-left panel in Figure 4.6 shows a highly nonlinear "delta" function, which is significantly better approximated by the diffusion wavelet bases (second panel, top) as compared to the Fourier (eigenfunction) basis (third panel, top).

The bottom panel shows that the difference between eigenfunctions and wavelet bases is much less pronounced for smooth functions.

4.5   BIBLIOGRAPHICAL REMARKS

A more detailed overview of diffusion wavelets is given in [30, 21], as part of a special issue of the journal *Applied and Computational Harmonic Analysis* (ACHA) on diffusion analysis. The description of diffusion wavelets in this chapter is based on [73].

CHAPTER 5

Scaling to Large Spaces

While harmonic analysis is a theoretically attractive framework for representation discovery, it can be computationally intractable to apply the framework to large discrete or continuous spaces. In this chapter, we describe ways of scaling harmonic analysis to continuous and large discrete spaces. We investigate several approaches, ranging from exploiting the structure of highly symmetric graphs [31, 61], to the use of sparsification and sampling methods [41] to streamline matrix computations. In addition, there has been a recent breakthrough in solving very large systems of linear Laplacian equations of the form $Lx = b$ [108], with asymptotic complexity $O(n \log^{O(1)} n)$. We see how to construct basis functions in irregular continuous domains called *manifolds* [70, 99], which are sets embedded in Euclidean spaces. Here, we are faced with a new challenge: we only have access to *samples* of the underlying manifold, and have to deal with an *out-of-sample* extension problem.

5.1 KRONECKER SUM DECOMPOSITION

We first analyze structured graphs that are constructed from simpler graphs, based on the notion of a *Kronecker product* [31]. We describe a general framework for scaling basis construction through Fourier and wavelet analysis to large *factored* discrete spaces using properties of product spaces, such as grids, cylinders, and tori. A crucial property of the graph Laplacian is that its embeddings are highly regular for structured graphs (see Figure 5.2). We will explain the reason for this property below, and how to exploit it to construct compact encodings of Laplacian bases. We should also distinguish the approach described in this section, which relies on an *exact* Kronecker decomposition of the Laplacian eigenspace in product spaces, with the *approximate* Kronecker decomposition described in the following section. The approach described here is applicable only to graphs which can be represented as the Kronecker sum of simpler graphs (this notion will be defined more precisely below, but it covers many standard graphs like grids). More generally, the weight matrices for arbitrary graphs can also be decomposed, although using the Kronecker *product*, where, however, the factorization is an approximation.

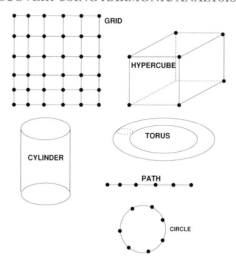

FIGURE 5.1: The spectrum and eigenspace of structured state spaces, including grids, hypercubes, cylinders, and tori, can be efficiently computed from "building block" subgraphs, such as paths and circles. This hierarchical framework greatly reduces the computational expense of computing and storing basis functions.

5.1.1 Product Spaces: Complex Graphs from Simple Ones

Building on the theory of graph spectra [31], we now describe a hierarchical framework for efficiently computing and compactly storing basis functions on product graphs. Many applications lead to *factored* representations where the space is generated as the Cartesian product of the values of variables. Consider a hypercube graph with d dimensions, where each dimension can take on k values. The size of the resulting graph is $O(k^d)$, and the size of each function on the graph is $O(k^d)$. Using the hierarchical framework presented below, the hypercube can be viewed as the *Kronecker sum* of d path or chain graphs, each of whose transition matrix is of size (in the worst case) $O(k^2)$. Now, each factored function can be stored in space $O(dk^2)$, and the cost of spectral analysis greatly reduces as well. Even greater savings can be accrued since usually only a small number of basis functions are needed relative to the size of the graph. Figure 5.1 illustrates the idea of scaling Fourier and wavelet basis functions to large product graphs.

Various compositional schemes can be defined for constructing complex graphs from simpler graphs [31]. We focus on compositions that involve the Kronecker (or the tensor) sum of graphs. Let $G_1, \ldots, G_n$ be n undirected graphs whose corresponding vertex and edge sets are specified as $G_i = (V_i, E_i)$. The *Kronecker sum graph* $G = G_1 \oplus \cdots \oplus G_n$ has the vertex set $V = V_1 \times \cdots \times V_n$, and edge set $E(u, v) = 1$, where $u = (u_1, \ldots, u_n)$ and $v = (v_1, \ldots, v_n)$, if and only if u_k is adjacent to v_k for some $u_k, v_k \in V_k$ and all $u_i = v_i$, $i \neq k$. For example, the grid graph illustrated in Figure 5.1 is the *Kronecker sum* of two path graphs; the hypercube is the Kronecker sum of three or more path graphs.

The Kronecker sum graph can also be defined using operations on the component adjacency matrices. If A_1 is a (p, q) matrix and A_2 is a (r, s) matrix, the Kronecker product matrix[1] $A = A_1 \otimes A_2$ is a (pr, qs) matrix, where $A(i, j) = A_1(i, j) * A_2$. In other words, each entry of A_1 is replaced by the product of that entry with the entire A_2 matrix. The Kronecker sum of two graphs $G = G_1 \oplus G_2$ can be defined as the graph whose adjacency matrix is the Kronecker sum $A = A_1 \otimes I_2 + A_2 \otimes I_1$, where I_1 and I_2 are the identity matrices of size equal to number of rows (or columns) of A_1 and A_2, respectively. The main result that we will exploit is that the eigenvectors of the Kronecker product of two matrices can be expressed as the Kronecker products of the eigenvectors of the component matrices.

Theorem 5.1. *Let A and B be full rank square matrices of size $r \times r$ and $s \times s$, respectively, whose eigenvectors and eigenvalues can be written as*

$$Au_i = \lambda_i u_i, \ 1 \le i \le r, \qquad Bv_j = \mu_j v_j, \ 1 \le j \le s.$$

Then, the eigenvalues and eigenvectors of the Kronecker product $A \otimes B$ and the Kronecker sum $A \oplus B$ are given as

$$(A \otimes B)(u_i \otimes v_j) = \lambda_i \mu_j (u_i \otimes v_j),$$
$$(A \oplus B)(u_i \otimes v_j) = (A \otimes I_s + I_r \otimes B)(u_i \otimes v_j) = (\lambda_i + \mu_j)(u_i \otimes v_j).$$

The proof of this theorem relies on the following identity regarding Kronecker products of matrices: $(A \otimes B)(C \otimes D) = (AC) \otimes (BD)$ for any set of matrices where the products AC and BD are well defined. We denote the set of eigenvectors of an operator T by the notation $X(T)$ and its spectrum by $\Sigma(T)$. A standard result that follows from the above theorem shows that the combinatorial graph Laplacian of a Kronecker sum of two graphs can be computed from the Laplacian of each subgraph. In contrast, the normalized Laplacian is not well-defined under sum, but has a well-defined semantics for the Kronecker or direct product of two graphs. The Kronecker product can also be used as a general method to approximate any matrix by factorizing it into the product of smaller matrices [120].

Theorem 5.2. *If $L_1 = L(G_1)$ and $L_2 = L(G_2)$ are the combinatorial Laplacians of graphs $G_1 = (V_1, E_1, W_1)$ and $G_2 = (V_2, E_2, W_2)$, then the spectral structure of the combinatorial Laplacian $L(G)$ of the Kronecker sum of these graphs $G = G_1 \oplus G_2$ can be computed as*

$$(\Sigma(L), X(L)) = \{\lambda_i + \kappa_j, l_i \otimes k_j\}, \ 1 \le i \le |V_1|, 1 \le j \le |V_2|,$$

---

[1]The Kronecker product of two matrices is often also referred to as the *tensor product*.

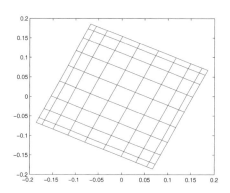
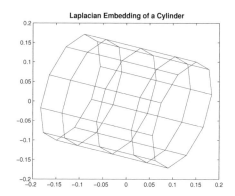

FIGURE 5.2: Left: this figure shows an embedding in $\mathbb{R}^2$ of a 10×10 grid world environment using "low-frequency" (smoothest) eigenvectors of the combinatorial Laplacian, specifically those corresponding to the second and third smallest eigenvalues. Right: the embedding of a "cylinder" graph using two low-order eigenvectors (third and fourth) of the combinatorial Laplacian. The cylinder graph is the Kronecker sum of a closed and open chain graph.

where λ_i is the ith eigenvalue of L_1 with the associated eigenvector l_i and κ_j is the jth eigenvalue of L_2 with the associated eigenvector k_j.

The proof is omitted, but fairly straightforward by exploiting the property that the Laplace operator acts on a function by summing the difference of its value at a vertex with those at adjacent vertices. Figure 5.2 illustrates this theorem, showing that the eigenvectors of the combinatorial Laplacian produce a regular embedding of a grid in 2D as well as a cylinder in 3D. These figures were generated as follows. For the grid shown on the left, the eigenvectors were generated as the Kronecker product of the eigenvectors of the combinatorial Laplacian for two chains of size 10. The figure shows the embedding of the grid graph where each state was embedded in $\mathbb{R}^2$ using the second and third smallest eigenvectors. For the cylinder on the right, the eigenvectors were generated as the Kronecker product of the eigenvectors of the combinatorial Laplacian for a ten-state closed chain and a five-state open chain. The embedding of the cylinder shown on the right was produced using the third and fourth eigenvector of the combinatorial Laplacian.

For the combinatorial Laplacian, the constant vector **1** is an eigenvector with associated eigenvalue $\lambda_0 = 0$. Since the eigenvalues of the Kronecker sum graph are the sums of the eigenvalues of the individual graphs, 0 will be an eigenvalue of the Laplacian of the sum graph as well. Furthermore, for each eigenvector v_i, the Kronecker product $v_i \otimes \mathbf{1}$ will also be an eigenvector of the sum graph. One consequence of these properties is that geometry is well preserved, so for example the combinatorial Laplacian produces well-defined embeddings of structured spaces.

5.2   SCALING TO LARGE GRAPHS USING APPROXIMATION METHODS

A variety of other approximation methods can be used to scale basis construction to large graphs, including matrix sparsification [1], low-rank approximation [46], graph partitioning [60], and Kronecker product approximation [120]. We review the latter two methods first, and then discuss the former methods in the context of approximating functions on manifolds.

5.2.1   Kronecker Product Approximation

The Kronecker product approximation [120] constructs two smaller stochastic matrices B and C whose Kronecker product $B \otimes C$ approximates a given matrix A. It is important to distinguish this approach from the Kronecker *sum* decomposition approach described in Section 5.1, where the factorization was not an approximation, but an exact decomposition assuming the overall state space was a product space. Kronecker product factorization can be applied to arbitrary weight matrices, but the decomposition is an approximation.

Let $P_r = D^{-1}W$ denote the random walk matrix, as described in Chapter 3. P_r can be approximated by a Kronecker product of two smaller stochastic matrices P_a and P_b, which minimizes the Fröbenius norm of the error:

$$f(P_a, P_b) = \min_{P_a, P_b} \left(\| P_r - P_a \otimes P_b \|_F \right).$$

Pitsianis [120] describes a separable least-squares method to decompose stochastic matrices, but one problem with this approach is that the decomposed matrices, although stochastic, are not guaranteed to be diagonalizable. This problem was addressed by Johns et al. [56], who applied this approach for learning to solve Markov decision processes (this application is described in more detail in Chapter 6).

To ensure the diagonalizability of the decomposed matrices, Johns et al. [56] incorporated an additional step using the Metropolis–Hastings algorithm [15] to approximate the smaller matrices P_a and P_b by *reversible matrices* P_a^r and P_b^r. Then, the eigenvectors of the original random walk matrix P_r can be approximated as the Kronecker product of the eigenvectors of the factorized smaller reversible matrices P_a^r and P_b^r (since the smaller matrices are reversible, they can also be symmetrized using the normalized Laplacian, which makes the numerical task of computing their eigenvectors much simpler). Using this approach, Johns et al. [56] were able to reduce the size of the random walk weight matrices by a significant amount compared to the full matrix. For example, in one problem, the original basis matrix was compressed by a factor of $36 : 1$, without significant loss in the solution quality. An important point to emphasize is that the full basis matrix never needs to be stored or computed in constructing the vertex embeddings from the smaller matrices. The factorization can be carried out recursively as well, leading to a further reduction in the size of the basis matrices.

5.2.2   Graph Partitioning and Fast Laplacian Solvers

A general divide-and-conquer strategy is to decompose the original graph into subgraphs, and then compute local basis functions on each subgraph. This strategy can be used on any graph, however unlike the methods described above, few theoretical guarantees can be provided except in special circumstances. A number of graph partitioning methods are available, including spectral methods that use the low-order eigenvectors of the Laplacian to decompose graphs [86], as well as hybrid methods that combine spectral analysis with other techniques. In Section 3.4, we described the connection between the eigenvectors of the graph Laplacian and graph partitioning.

Graph partitioning is a well-studied topic, and there are a large variety of non-spectral methods as well. METIS [60] is a fast graph partitioning algorithm that can decompose even very large graphs on the order of 10^6 vertices. METIS uses a *multiscale* approach to graph partitioning, where the original graph is "coarsened" by collapsing vertices (and their associated edges) to produce a series of smaller graphs, which are successively partitioned followed by uncoarsening steps mapping the partitions found back to the lower-level graphs. We will illustrate the use of METIS in Chapter 7, where it is used to decompose large 3D models.

There has been a recent breakthrough in solving systems of linear equations of the form $Lx = b$, where L is a graph Laplacian. In particular, Spielman and Teng [108] showed that general Laplacian systems of linear equations can be solved in time $O(n \log^{O(1)} n)$, and Laplacians corresponding to planar graphs can be solved in time $O(n \log^3 n)$. Similarly, Koutis and Miller [64] describe a parallel $O(n^{1/6})$ algorithm for solving linear systems of Laplacian equations on planar graphs. Collectively, these algorithms rely on sophisticated methods for partitioning large graphs. Another area of research into fast methods for solving symmetric systems of linear equations, such as the Laplacian systems, is *algebraic multigrid methods* [20].

5.3   SCALING TO CONTINUOUS SPACES

Thus far, we have restricted our attention to Fourier and wavelet analysis in discrete spaces, or in continuous spaces where the basis functions were already predefined. We now discuss the use of harmonic analysis to construct basis functions in irregular continuous domains called *manifolds* [70, 99], which usually refer to sets embedded in Euclidean spaces. We are faced with a new challenge: we only have access to *samples* of the underlying manifold, and need to extrapolate these samples to new points during the testing phase when the newly discovered basis functions are being applied. For example, in a control setting where an agent is learning to solve a planning problem [111], the agent will encounter new situations that were not previously observed during the process of basis construction. This problem is generally referred to as the *out-of-sample* extension problem [90].

To solve the out-of-sample extension problem, we introduce a popular method called the *Nyström* extension [7, 41], which allows a principled approach to extrapolating sample values of eigenfunctions to new points. We show how the Nyström interpolation can be viewed as doing a low-rank approximation of a positive semi-definite (kernel) matrix.

5.4 RIEMANNIAN MANIFOLDS

This section briefly introduces the Laplace–Beltrami operator in the general setting of Riemannian manifolds [99], building on the intuitions gained in the more familiar setting of graphs [26]. Riemannian manifolds have been actively studied recently in machine learning in several contexts. It has been known for over 50 years that the space of probability distributions forms a Riemannian manifold, with the Fisher information metric representing the Riemann metric on the tangent space. This observation has been applied to design new types of kernels for supervised machine learning [66] and faster policy gradient methods using the natural Riemannian gradient on a space of parametric policies [6, 58, 94]. One popular application of manifold learning is *semi-supervised learning* [10], where a large set of *unlabeled* points are used to extract a representation of the underlying manifold and improve classification accuracy. The Laplacian on Riemannian manifolds and its eigenfunctions [99], which form an orthonormal basis for square-integrable functions on the manifold (Hodge's theorem), generalize Fourier analysis to manifolds. Historically, manifolds have been applied to many problems in AI, for example configuration space planning in robotics, but these problems assume a model of the manifold is known [68, 69], unlike here where only samples of a manifold are given. Recently, there has been rapidly growing interest in manifold learning methods, including ISOMAP [114], LLE [101], and Laplacian eigenmaps [10]. These methods have been applied to nonlinear dimensionality reduction as well as semi-supervised learning on graphs [10, 28, 128].

5.4.1 Manifolds

This section introduces the Laplace–Beltrami operator in the general setting of Riemannian manifolds [99], as an extension of the graph Laplacian operator described earlier in the more familiar setting of graphs [26].

Formally, a *manifold* $\mathcal{M}$ is a *locally Euclidean* set, with a *homeomorphism* (a bijective or one-to-one and onto mapping) from any open set containing an element $p \in \mathcal{M}$ to the n-dimensional Euclidean space $\mathbb{R}^n$. Manifolds with *boundaries* are defined using a homeomorphism that maps elements to the upper half plane $\mathcal{H}^n$ [70]. A manifold is a topological space, i.e. a collection of open sets closed under finite intersection and arbitrary union. In smooth manifolds, the homeomorphism becomes a *diffeomorphism*, or a continuous bijective mapping with a continuous inverse mapping, to the Euclidean space $\mathbb{R}^n$.

In a smooth manifold, a diffeomorphism mapping any point $p \in M$ to its *coordinates* $(\rho_1(p), \ldots, \rho_n(p))$ should be a differentiable function with a differentiable inverse. Given two coordinate functions $\rho(p)$ and $\xi(p)$, or *charts*, the induced mapping $\psi : \rho \circ \xi^{-1} : \mathbb{R}^n \to \mathbb{R}^n$ must have continuous partial derivatives of all orders. *Riemannian* manifolds are smooth manifolds where the Riemann metric defines the notion of length. Given any element $p \in M$, the *tangent space* $T_p(M)$ is an n-dimensional vector space that is isomorphic to $\mathbb{R}^n$. A Riemannian manifold is a smooth manifold M with a family of smoothly varying positive semi-definite inner products g_p, $p \in M$ where $g_p : T_p(M) \times T_p(M) \to \mathbb{R}$. For the Euclidean space $\mathbb{R}^n$, the tangent space $T_p(M)$ is clearly isomorphic to $\mathbb{R}^n$ itself. One example of a Riemannian inner product on $\mathbb{R}^n$ is simply $g(x, y) = \langle x, y \rangle_{\mathbb{R}^n} = \sum_i x_i y_i$, which remains the same over the entire space. If the space is defined by the set of probability distributions $P(X|\theta)$, then one example of a Riemann metric is given by the Fisher information metric $\mathcal{I}(\theta)$ [66].

5.4.2   Hodge Theorem

Hodge's theorem [99] states that any smooth function on a compact manifold has a discrete spectrum mirrored by the *eigenfunctions* of Δ, the Laplace–Beltrami self-adjoint operator. On the manifold $\mathbb{R}^n$, the Laplace–Beltrami operator is $\Delta = \sum_i \frac{\partial^2}{\partial x_i^2}$ (often written with a $-$ sign for convention). Functions that solve the equation $\Delta f = 0$ are called *harmonic functions* [5]. For example, on the plane $\mathbb{R}^2$, the "saddle" function $x^2 - y^2$ is harmonic. *Eigenfunctions* of Δ are functions f such that $\Delta f = \lambda f$, where λ is an eigenvalue of Δ. If the domain is the unit circle S^1, the trigonometric functions $\sin(\theta)$ and $\cos(\theta)$ form eigenfunctions, which leads to *Fourier* analysis. Abstract harmonic analysis generalizes Fourier methods to smooth functions on arbitrary Riemannian manifolds. The *smoothness functional* for an arbitrary real-valued function on the manifold $f : M \to \mathbb{R}$ is given by

$$S(f) \equiv \int_M |\nabla f|^2 \, d\mu = \int_M f \Delta f d\mu = \langle \Delta f, f \rangle_{\mathbb{L}^2(M)},$$

where $\mathbb{L}^2(M)$ is the space of smooth functions on M, and ∇f is the gradient vector field of f. We refer the reader to [99] for an introduction to the Riemannian geometry and properties of the Laplacian on Riemannian manifolds. Let (M, g) be a smooth compact connected Riemannian manifold. The Laplacian is defined as

$$\Delta = \text{div grad} = \frac{1}{\sqrt{\det g}} \sum_{ij} \partial_i \left(\sqrt{\det g} \ g^{ij} \partial_j \right),$$

where g is the Riemannian metric, $\det g$ is the measure of volume on the manifold, ∂_i denotes differentiation with respect to the ith coordinate function, and div and grad are the Riemannian divergence and gradient operators, respectively. We say that $\phi : \mathcal{M} \to \mathbb{R}$ is an eigenfunction of Δ if $\phi \neq 0$ and there exists $\lambda \in \mathbb{R}$ such that

$$\Delta\phi = \lambda\phi .$$

If $\mathcal{M}$ has a boundary, special conditions need to be imposed. Typical boundary conditions include Dirichlet conditions, enforcing $\phi = 0$ on $\partial\mathcal{M}$ and Neumann conditions, enforcing $\partial_\nu\phi = 0$, where ν is the normal to $\partial\mathcal{M}$. The set of λ's for which there exists an eigenfunction is called the spectrum of Δ, and is denoted by $\sigma(\Delta)$. We always consider eigenfunctions which have been $\mathbb{L}^2$-normalized, i.e. $||\phi||_{\mathbb{L}^2(\mathcal{M})} = 1$.

The quadratic form associated with the Laplacian is the Dirichlet integral

$$S(f) := \int_{\mathcal{M}} ||\mathrm{grad}\, f||^2 \mathrm{d\,vol} = \int_{\mathcal{M}} f\Delta f \mathrm{d\,vol} = \langle \Delta f, f \rangle_{\mathbb{L}^2(\mathcal{M})} = ||\mathrm{grad}\, f||_{\mathbb{L}^2(\mathcal{M})},$$

where $\mathbb{L}^2(\mathcal{M})$ is the space of square-integrable functions on $\mathcal{M}$, with respect to the natural Riemannian volume measure. It is natural to consider the space of functions $\mathcal{H}^1(\mathcal{M})$ defined as follows:

$$\mathcal{H}^1(\mathcal{M}) = \left\{ f \in \mathbb{L}^2(\mathcal{M}) : ||f||_{\mathcal{H}^1(\mathcal{M})} := ||f||_{\mathbb{L}^2(\mathcal{M})} + S(f) \right\} . \tag{5.1}$$

So clearly $\mathcal{H}^1(\mathcal{M}) \subsetneq \mathbb{L}^2(\mathcal{M})$ since functions in $\mathcal{H}^1(\mathcal{M})$ have a square-integrable gradient. The smaller the $\mathcal{H}^1$-norm of a function, the "smoother" the function is, since it needs to have small gradient. Observe that if ϕ_λ is an eigenfunction of Δ with eigenvalue λ, then $S(\phi_\lambda) = \lambda$: the larger is λ, the larger the square-norm of the gradient of the corresponding eigenfunction, i.e. the more oscillating the eigenfunction is.

Theorem 5.3 (Hodge [99]). *Let $(\mathcal{M}, g)$ be a smooth compact connected oriented Riemannian manifold. The spectrum $0 \leq \lambda_0 \leq \lambda_1 \leq \cdots \leq \lambda_k \leq \cdots$, $\lambda_k \to +\infty$, of Δ is discrete, and the corresponding eigenfunctions $\{\phi_k\}_{k\geq 0}$ form an orthonormal basis for $\mathbb{L}^2(\mathcal{M})$.*

In other words, Hodge's theorem shows that a smooth function $f \in \mathbb{L}^2(\mathcal{M})$ can be expressed as $f(x) = \sum_{i=0}^{\infty} a_i e_i(x)$, where e_i are the eigenfunctions of Δ, i.e. $\Delta e_i = \lambda_i e_i$. The smoothness $S(e_i) = \langle \Delta e_i, e_i \rangle_{\mathbb{L}^2(\mathcal{M})} = \lambda_i$. In particular, any function $f \in \mathbb{L}^2(\mathcal{M})$ can be expressed as $f(x) = \sum_{k=0}^{\infty} \langle f, \phi_k \rangle \phi_k(x)$, with convergence in $\mathbb{L}^2(\mathcal{M})$ [90].

5.5   THE NYSTRÖM INTERPOLATION OF EIGENFUNCTIONS

To learn functions on manifolds, it is necessary to be able to extend eigenfunctions computed on a set of points $\in \mathbb{R}^d$ to new unseen points. We describe here the Nyström method, which can be combined with iterative updates and randomized algorithms for low-rank approximations. The Nyström method interpolates the value of eigenvectors computed on sample states to novel states, and is an application of a classical method used in the numerical solution of integral equations [7]. It can be viewed as a technique for approximating a positive semi-definite matrix from a low-rank approximation. In this context it can be related to randomized algorithms for low-rank approximation of large matrices [46]. Let us review the Nyström method in its basic form. Suppose we have a positive semi-definite operator K, with rows and columns indexed by some measure space $(\mathbb{X}, \mu)$. K acts on a vector space of functions on X by the formula

$$Kf(x) = \int_\mathbb{X} K(x, y)f(y)\mathrm{d}\mu(y), \tag{5.2}$$

for f in some function space on $\mathbb{X}$. Examples include

(i) $\mathbb{X} = \{0, \ldots, n\}$, μ assigns mass 1 to each element of $\mathbb{X}$, then K is an $n \times n$ matrix acting on n-dimensional vectors by matrix multiplication on the left.

(ii) $\mathbb{X} = \mathbb{R}$, μ is the Lebesgue measure, $K_\sigma(x, y) = \mathrm{e}^{-\frac{|x-y|^2}{\sigma}}$, and K acts on square-integral functions f on $\mathbb{R}$ by $K_\sigma f(x) = \int_{-\infty}^{+\infty} \mathrm{e}^{-\frac{|x-y|^2}{\sigma}} f(y)\mathrm{d}y = K_\sigma * f$.

(iii) $\mathbb{X}$ is a compact Riemannian manifold $(\mathcal{M}, \rho)$ equipped with the measure corresponding to the Riemannian volume, Δ is the Laplace–Beltrami operator on $\mathcal{M}$, with Dirichlet or Neumann boundary conditions if $\mathcal{M}$ has a boundary, and $K = (I - \Delta)^{-1}$ is the Green's function or potential operator associated with Δ.

Since K is positive semi-definite, by the spectral theorem (described in Chapter 2 for the finite-dimensional case), it has a square root F, i.e. $K = F^T F$. Sometimes this property is expressed by saying that K is a Gram matrix (see Section 2.6.3), since we can interpret $K(x, y)$ as the inner product between the xth and yth columns of F. In applications, operators on uncountable spaces (such as $\mathbb{R}$ or a manifold $\mathcal{M}$ as in the examples above) are approximated by a finite discretization $x_1, \ldots, x_n$, in which case $\mathbb{X} = \{0, \ldots, n\}$, the measure μ is an appropriate set of weights on the n points, and K is a $n \times n$ matrix acting on n-dimensional vectors. To simplify the notation we use this discrete setting in what follows.

The Nyström approximation starts with a choice of a partition of the columns of F into two subsets F_1 and F_2. Let k be the cardinality of F_1, so that F_1 can be represented as $n \times k$ matrix and F_2 as a $n \times (n - k)$ matrix. One can then write

$$K = \begin{pmatrix} F_1^T F_1 & F_1^T F_2 \\ F_2^T F_1 & F_2^T F_2 \end{pmatrix}.$$

The Nyström method consists of the approximation

$$F_2^T F_2 \sim (F_1^T F_2)^T (F_1^T F_1)^{-1} (F_1^T F_2).\tag{5.3}$$

The quantity on the right-hand side requires only the knowledge of $(F_1^T F_2)$ and $F_1^T F_1$, i.e. the first k rows (or columns) of K. Moreover if the matrix K has rank k and F_1 spans the range of K, then the Nyström approximation is in fact exactly equal to $F_2^T F_2$.

This technique applies to the discretization of integral equations [7], where the k points F_1 can be chosen according to a careful mathematical and numerical analysis of the problem, and has been applied to speeding up the computations in learning and clustering algorithms [96, 126, 12]. The natural question that arises is of course how to choose F_1 in these situations. Various heuristics exist, and mixed results have been obtained [96]. The most desirable choice of F_1, when the error of approximation is measured by $||F_2^T F_2 - (F_1^T F_2)^T (F_1^T F_1)^{-1} (F_1^T F_2)||_2$ (or, equivalently, the Fröbenius norm), would be to pick F_1 such that its span is as close as possible to the span of the top k singular vectors of K. Several numerical algorithms exist, which in general require $\mathcal{O}(kN^2)$ computations. One can use randomized algorithms, which pick rows (or columns) of K accordingly to some probability distribution (e.g. dependent on the norm of the row or column). There are guarantees that these algorithms will select with high probability a set of rows whose span is close to that of the top singular vectors: see for example [40, 41, 46].

To learn functions on continuous manifolds, it is necessary to be able to extend eigenfunctions computed on a set of points $\in \mathbb{R}^n$ to new unexplored points. We describe here a special instance of the Nyström method for the normalized Laplacian described in Chapter 3. We begin by restating Equation (5.2) for eigenfunctions ϕ of the kernel K:

$$\int_D K(x, y)\phi(y)\mathrm{d}y = \lambda\phi(x), \quad \forall x \in D,\tag{5.4}$$

where D can be any domain, e.g. $\mathbb{R}$. Using the standard quadrature approximation, the above integral can be written as

$$\int_D K(x, y)\phi(y)\mathrm{d}y \approx \sum_{i=1}^{n} w_i k(x, s_i)\hat{\phi}(s_i),\tag{5.5}$$

where w_i are the quadrature weights, s_i are n selected sample points, and $\hat{\phi}$ is an approximation to the true eigenfunction. Combining Equations (5.4) and (5.5) gives us

$$\sum_{i=1}^{n} w_i k(x, s_i)\hat{\phi}(s_i) = \hat{\lambda}\hat{\phi}(x).\tag{5.6}$$

By letting x denote any set of n points, for example the set of quadrature points s_i itself, the kernel $k(s_i, s_j)$ becomes a symmetric matrix. This enables computing the approximate eigenfunction at any new point as

$$\hat{\phi}_m(x) = \frac{1}{\hat{\lambda}} \sum_{i=1}^n w_i k(x, s_i) \hat{\phi}_m(s_i). \tag{5.7}$$

Let us instantiate Equation (5.7) in the context of the normalized Laplacian $\mathcal{L} = I - D^{-\frac{1}{2}} W D^{-\frac{1}{2}}$. First, note that if λ_i is an eigenvalue of $\mathcal{L}$, then $1 - \lambda_i$ is the corresponding eigenvalue of the diffusion matrix $D^{-\frac{1}{2}} W D^{-\frac{1}{2}}$. Applying the Nyström extension for computing the eigenfunctions of the normalized Laplacian $\mathcal{L}\phi_i = \lambda_i \phi_i$, we get the equation

$$\phi_i(x) = \frac{1}{1 - \lambda_i} \sum_{y \sim x} \frac{w(x, y)}{\sqrt{d(x)d(y)}} \phi_i(y), \tag{5.8}$$

where $d(z) = \sum_{y \sim z} w(z, y)$, and x is a new vertex in the graph. Note that the weights $w(x, y)$ from the new state x to its nearest neighbors y in the previously stored samples is determined at "run time" using the same nearest-neighbor weighting algorithm used to compute the original weight matrix W. An extensive discussion of the Nyström method is given in [41], and more details of its application to learning control in MDPs are given in Chapter 6 as well as in [80].

Figure 5.3 illustrates the basic idea. Note that the Nyström method does *not* require recalculating eigenvectors—in essence, the embedding of a new state is computed by averaging over the already computed embeddings of "nearby" states. In practice, significant speedups can be exploited by using the following optimizations. Once the bases are defined over a sampled set of points, the Nyström extended embeddings of the remaining training samples needs to be calculated only once, and henceforth can be cached. During testing, the Nyström embeddings of novel points encountered must be computed, but since the eigenvectors are defined over a relatively small core set of sample states, the extensions can be computed very efficiently using a fast nearest-neighbor algorithm.

The Nyström method can be refined with fast iterative updates as follows: first compute an extension of the eigenvectors to new points (states), to obtain approximated eigenvectors of the extended graph $\{\tilde{\phi}_i\}$. Input these eigenvectors into an iterative eigensolver as initial approximate eigenvectors: after very few iterations the eigensolver will refine these initial approximate eigenvectors into more precise eigenvectors on the larger graph. The extra cost of this computation is $\mathcal{O}(IN)$ if I iterations are necessary, and if the adjacency matrix of the extended graph is sparse (only N nonzero entries).

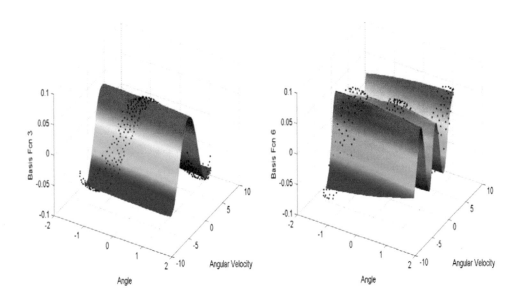

FIGURE 5.3: This figure illustrates the Nyström interpolation method for extending eigenfunctions on samples to new states. Left: the third eigenvector of the Laplacian plotted on a set of samples (shown as filled dots) drawn from a random walk in the inverted pendulum domain, as well as its Nyström interpolated values. Right: the Nyström interpolated sixth eigenvector illustrated on the entire state space as well as on the actual samples (again shown as filled dots).

5.6   SAMPLING TECHNIQUES

The smoothness of functions on a manifold as defined by Equation (5.1) determines the number of samples necessary to approximate the function up to a given precision. This number of samples is independent of the number of points explored. Consider the following simple example.

Example 5.1.   Suppose the state space is the interval $[0, 1]$, and that the function f is band-limited with bandwidth B. This means that the Fourier transform $\hat{f}$ is supported in $[-B, B]$. Then by the Whittaker–Shannon sampling theorem [81], only $B/(2\pi)$ equispaced samples are needed to recover V *exactly*.

Suppose we have observed samples $\mathcal{S}'$ in a space $\mathcal{S}$, and that the function f is smooth so that a subset $\mathcal{S}''$ much smaller than $\mathcal{S}'$ would suffice to determine f. We propose two simple methods in order to select $\mathcal{S}''$.

Purely random subsampling.   We fix $|\mathcal{S}''|$, and select $|\mathcal{S}''|$ points uniformly at random in $\mathcal{S}'$. For very large $|\mathcal{S}'|$ one would expect that the points in $\mathcal{S}''$ are going to be well-spread in $\mathcal{S}'$.

Well-spread random net. The previous algorithm has two main drawbacks: first, it is not clear how to select $|\mathcal{S}''|$, even if in theory this number can be determined by knowing the complexity of the function to be approximated. Second, the points in $\mathcal{S}''$ are not going to be necessarily well-spread in $\mathcal{S}'$: while it is true that for large $|\mathcal{S}'|$, with very high probability, no two points in $\mathcal{S}''$ are going to be very close, it is not true that the points in $\mathcal{S}''$ are going to be roughly equidistant nor well equidistributed in balls contained in $\mathcal{S}'$.

In order to guarantee that the set of points is well-spread, we consider the following construction. We define an ϵ-net of points in $\mathcal{S}'$ to be a subset $\mathcal{S}''$ such that no two points are closer than ϵ, and that for every point y in $\mathcal{S}'$, there is a point in $\mathcal{S}''$ which is not farther than ϵ from y. One can construct a (random) ϵ-net in $\mathcal{S}'$ as follows. Pick $x_0 \in \mathcal{S}'$ at random. By induction, for $k \geq 1$ suppose $x_0, x_1, \ldots, x_k$ have been picked so that the distance between any pair is larger than ϵ. If

$$R_k := \mathcal{S}' \setminus (\cup_{l=1}^{k} B_\epsilon(x_l)) \,,$$

is empty, stop, otherwise pick a point x_{k+1} in R_k. By definition of R_k the distance between x_{k+1} and any of the points $x_0, \ldots, x_k$ is not smaller than ϵ. When this process stops, say after k^* points have been selected, for any $y \in \mathcal{S}'$ we can find a point in $\mathcal{S}''$ not farther than ϵ, for otherwise $y \in R_{k^*}$ and the process would not have stopped. One can prove upper bounds of the distance between the eigenfunctions of the Laplacian on $\mathcal{S}'$ and the eigenfunctions of the Laplacian on $\mathcal{S}''$, which depend on ϵ and the order of the eigenfunction.

Convergence: From Graph Laplacian to Manifold Laplacian. Theoretical guarantees on the convergence of the graph Laplacian, described in Chapter 3, to the manifold Laplacian described in this chapter have been investigated. Belkin and Niyogi [11] and Hein et al. [52] study sampling conditions under which the various graph Laplacians converge to the Laplace–Beltrami operator on the underlying manifold. For example, Hein et al. [52] show that under non-uniform sampling conditions, the random walk Laplacian $L_r = I - D^{-1}W$ converges to a weighted Laplace–Beltrami operator.

5.7 EXPLOITING DOMAIN KNOWLEDGE

A general way to scale basis construction methods is to exploit available domain knowledge, for example, knowledge of the global geometry of the underlying space. Chapter 6 illustrates several examples of how domain knowledge can be used to accelerate basis construction for solving Markov decision processes, including knowledge of the state geometry, knowledge of distance metrics on the space, and finally task-specific information in the form of rewards. Chapter 9 shows how to exploit the properties of a symmetry group acting on the underlying space to scale basis function construction.

5.8   BIBLIOGRAPHICAL REMARKS

The Kronecker sum and product decompositions of graphs are described in detail in [31]. Section 5.1 is based on [79]. Rosenberg [99] gives a highly mathematical treatment of the Laplacian on a Riemannian manifold. Belkin and Niyogi [91] pioneered the study of the Laplacian in machine learning, originally in the context of semi-supervised learning. Section 5.5 and Figure 5.3 is based on [79]. Section 5.6 is based on [73].

CHAPTER 6

Case Study: State-Space Planning

In this chapter, we describe a detailed case study of representation discovery applied to solve stochastic state-space planning problems [74, 75, 78, 80, 79]. Markov decision processes [98] have emerged as the standard mathematical framework to model sequential decision-making in a variety of areas, ranging from game-playing [116], manufacturing [33], robotics [77, 89], and scheduling [127]. Solving a Markov decision process requires computing and approximating *value* functions. Often, value functions are approximated in large spaces as linear combinations of pre-defined basis functions [14]. We show that harmonic analysis provides a powerful way to synthesize new basis functions that in some cases can outperform carefully hand-tuned basis functions in challenging control tasks [79]. These automatically generated basis functions are called "proto-value" functions (PVFs) [74], since they have global support on the state space similar to value functions, and all value functions on a given state space can be expressed within their span. This chapter also shows that diffusion wavelets can be used to generate compact proto-value functions, and are additionally useful in compressing powers of transition matrices, leading to a novel way of evaluating a fixed plan (or policy) in a Markov decision process [72].

6.1 INTRODUCTION

This chapter describes a novel *spectral* framework for solving Markov decision processes (MDPs) [98] where both the underlying representation or basis functions and (approximate) optimal policies within the (linear) span of these basis functions are simultaneously learned. This framework addresses a major open problem not addressed by much previous work in the field of *approximate dynamic programming* [14] and *reinforcement learning* [111], where the set of "features" or *basis functions* mapping a state s to a k-dimensional real vector $\phi(s) \in \mathbb{R}^k$ is usually hand-engineered.

The overall framework can be summarized briefly as follows. The underlying task environment is modeled as an MDP, where the system dynamics and reward function are typically assumed to be unknown. An agent explores the underlying state space by carrying out actions using some policy, say a random walk. The agent constructs a (directed or undirected) graph connecting states that are "nearby". In the simplest setting, the diffusion model is defined by the combinatorial graph Laplacian matrix $L = D - W$. Basis functions are derived by diagonalizing the Laplacian matrix L, specifically by finding its "smoothest" eigenvectors that correspond to the smallest eigenvalues. The similarity between value functions and the eigenvectors of the graph Laplacian sometimes can be remarkable, leading to a highly compact encoding (measured in terms of the number of basis functions needed to encode a value function). Laplacian basis functions can be used in conjunction with a standard "black box" parameter estimation method, such as Q-learning [125] or least-squares policy iteration (LSPI) [67] to find the best policy representable within the space of the chosen basis functions.

In many continuous control tasks, there are often physical constraints that limit the "degrees of freedom" to a lower-dimensional manifold, resulting in motion along highly constrained regions of the state space. Instead of placing basis functions uniformly in all regions of the state space, the proposed framework recovers the underlying manifold by building a graph based on the samples collected over a period of exploratory activity. The basis functions are then computed by diagonalizing a diffusion operator (the Laplacian) on the space of functions on the graph, and are thereby customized to the manifold represented by the state (action) space of a particular control task.

"Inaccessible" regions of the state space can be exploited in focusing the function approximator to accessible regions. Parametric approximators, as typically constructed, do not distinguish between accessible and inaccessible regions. The spectral approach goes beyond modeling just the reachable state space, in that it also models the local *non-uniformity* of a given region. This non-uniform modeling of the state space is facilitated by constructing a graph operator which models the local density across regions. By constructing basis functions adapted to the non-uniform density and geometry of the state space, the spectral approach extracts significant topological information from trajectories.

Dayan [35] proposed the idea of building *successor representations*. While this approach was restricted to policy evaluation in simple discrete MDPs, the idea of constructing representations that are faithful to the underlying dynamics of the MDP was a key motivation underlying this work. Drummond [42] used techniques from computer vision to detect nonlinearities in value functions. Finally, Foster and Dayan [45] attempt to find the "building blocks" of value functions by constructing a probabilistic generative (mixture) model using maximum likelihood estimation techniques. Proto-value functions can be viewed similarly as the building blocks of the set of value functions on a given state space, except that they are constructed non-parametrically.

6.2   MARKOV DECISION PROCESSES

A discrete Markov decision process (MDP) $M = (S, A, P^a_{ss'}, R^a_{ss'})$ is defined as a finite set of discrete states S, a finite set of actions A, a transition model $P^a_{ss'}$ specifying the distribution over future states s' when an action a is performed in state s, and a corresponding reward model

$R^a_{ss'}$ specifying a scalar cost or reward [98]. In continuous Markov decision processes, the set of states $\subseteq \mathbb{R}^d$. Abstractly, a value function is a mapping $S \to \mathbb{R}$ or equivalently (in discrete MDPs) a vector $\in \mathbb{R}^{|S|}$. Given a policy $\pi : S \to A$ mapping states to actions, its corresponding value function V^π specifies the expected long-term discounted sum of rewards received by the agent in any given state s when actions are chosen using the policy. Any optimal policy π^* defines the same unique optimal-value function V^*, which satisfies the nonlinear system of equations referred to as the "Bellman equations":

$$V^*(s) = \max_a \left(R_{sa} + \gamma \sum_{s' \in S} P^a_{ss'} V^*(s') \right), \tag{6.1}$$

where $R_{sa} = \sum_{s' \in s} P^a_{ss'} R^a_{ss'}$ is the expected immediate reward. Often, it is useful to summarize the above equation as the fixed point of an operator T^*, where $V^* = T^*(V^*)$. Value functions are mappings from the state space to the expected long-term discounted sum of rewards received by following a fixed (deterministic or stochastic) policy π. The value function V^π associated with following a (deterministic) policy π can be defined also as a (linear) Bellman equation:

$$V^\pi(s) = R_{s\pi(s)} + \gamma \sum_{s' \in S} P^{\pi(s)}_{ss'} V^\pi(s'). \tag{6.2}$$

Similarly, the above equation can be viewed as computing the fixed point of the operator T^π, where $V^\pi = T^\pi(V^\pi)$. It is also useful to define Bellman equations for *action-value* functions $Q^\pi(s, a)$, which represents the expected cumulative reward received for doing action a *once*, and thereafter following policy π:

$$Q^\pi(s, a) = R_{sa} + \gamma \sum_{s' \in S} P^a_{ss'} V^\pi(s'). \tag{6.3}$$

The optimal-value function $V^*(s) = \max_a Q^*(s, a)$. Value functions (or action-value functions) in an MDP are the long-term result of rewards " diffusing" through the state space governed by the underlying system dynamics matrix P^π. Let R^π be a (column) vector of size $|S|$ of rewards. The value function associated with policy π can also be expressed as the Neumann geometric series of powers of the transition matrix:

$$V^\pi = (I - \gamma P^\pi)^{-1} R^\pi = \left(I + \gamma P^\pi + \gamma^2 (P^\pi)^2 + \cdots \right) R^\pi. \tag{6.4}$$

The term $(I - \gamma P^\pi)^{-1}$ is generally referred to as the Green's function, particularly in the form $(I - T)^{-1}$ where T is a diffusion operator. Value functions generally satisfy two key properties: they are typically *smooth* (as quantified by the Sobolev norm introduced in Chapter 3), and they usually reflect the geometry of the environment. Smoothness derives from the fact that the value at a given state $V^\pi(s)$ is always a function of values at "neighboring" states. Consequently, it is natural to construct basis functions for approximating value functions that share these two properties.

6.2.1   Hilbert Space Formulation of Value Function Approximation

Let us define a set of *basis functions* $F_\Phi = \{\phi_1, \ldots, \phi_k\}$, where each basis function represents a "feature" $\phi_i : S \to \mathbb{R}$. The basis function matrix Φ is an $|S| \times k$ matrix, where each column is a particular basis function evaluated over the state space, and each row is the set of all possible basis functions evaluated on a particular state. Approximating a value function using the matrix Φ can be viewed as projecting the value function onto the column space spanned by the basis functions ϕ_i,

$$V^\pi \approx \hat{V}^\pi = \Phi w^\pi = \sum_i w_i^\pi \phi_i.$$

Mathematically speaking, this problem can be rigorously formulated using the framework of best approximation in inner product spaces [37]. In fact, it is easy to show that the space of value functions represents a Hilbert space, or a complete inner product space [121]. For simplicity, we focus on the simpler problem of approximating a fixed policy π, which defines a Markov chain where ρ^π represents its invariant (long-term) distribution. This distribution defines a Hilbert space, where the inner product is given by

$$\langle V_1, V_2 \rangle_\pi = \sum_{s \in S} V_1(s) V_2(s) \rho^\pi(s).$$

The "length" or norm in this inner product space is defined as $\|V\|_\pi = \sqrt{\langle V, V \rangle_\pi}$. Value function approximation can thus be formalized as a problem of best approximation in a Hilbert space [37]. It is well known (see Equation (2.5) for the general derivation) that if the basis functions ϕ_i are *orthonormal* (unit-length and mutually perpendicular), the best approximation of the value function V^π can be expressed by its projection onto the space spanned by the basis functions, or more formally

$$M_\Phi^\pi(V^\pi) = \sum_{i \in I} \langle V^\pi, \phi_i \rangle_\pi \, \phi_i,$$

where M_Φ^π is the projection operator, and I is the set of indices that define the basis set. In finite MDPs, the best approximation can be characterized using the weighted least-squares projection matrix

$$M_\Phi^\pi = \Phi(\Phi^T D_{\rho^\pi} \Phi)^{-1} \Phi^T D_{\rho^\pi},$$

where D_{ρ^π} is a diagonal matrix whose entries represent the distribution ρ^π. We know the Bellman operator T^π defined above has a fixed point $V^\pi = T^\pi(V^\pi)$. Many least-squares parameter estimation methods, including LSPI [67] and LSTD [19], can be viewed as finding the fixed point of the *combined* operator $M_\Phi^\pi T^\pi$

$$\hat{V}^\pi = \Phi w^\pi = M_\Phi^\pi (T^\pi(\Phi w^\pi)),$$

using a sequence of iterates $V^{m+1} = M_\Phi^\pi (T^\pi(V^m))$. How far is the fixed point of T^π, namely V^π, from the fixed point of $M_\Phi^\pi T^\pi$? To answer this question precisely, first it is useful to know that the operator T^π is a *contraction* mapping. Formally, an operator $\mathcal{F}$ on the Hilbert space of value functions is a contraction mapping if

$$\|\mathcal{F}V_1 - \mathcal{F}V_2\|_\pi \leq \beta \|V_1 - V_2\|_\pi,$$

where $\beta \in [0, 1)$ is the contraction factor. For any operator $\mathcal{F}$ with contraction factor β, we can prove a bound on the distance between the fixed point of $\mathcal{F}$, say $V_\mathcal{F}$, and the fixed point of the combined projection operator $M_\Phi \mathcal{F}$, denoted by $\hat{V}_\mathcal{F}^\Phi$. Here, M_Φ is the projection operator onto the basis spanned by F_Φ. This proof follows directly from the general properties of projection in Hilbert spaces given in Chapter 2:

$$
\begin{aligned}
\|V_\mathcal{F}^\Phi - V_\mathcal{F}\|^2 &= \|V_\mathcal{F}^\Phi - M_\Phi V_\mathcal{F} + M_\Phi V_\mathcal{F} - V_\mathcal{F}\|^2 \\
&= \|V_\mathcal{F}^\Phi - M_\Phi V_\mathcal{F}\|^2 + \|M_\Phi V_\mathcal{F} - V_\mathcal{F}\|^2 \\
&= \|M_\Phi \mathcal{F} V_\mathcal{F}^\Phi - M_\Phi \mathcal{F} V_\mathcal{F}\|^2 + \|M_\Phi V_\mathcal{F} - V_\mathcal{F}\|^2 \\
&\leq \kappa^2 \|V_\mathcal{F}^\Phi - V_\mathcal{F}\|^2 + \|M_\Phi V_\mathcal{F} - V_\mathcal{F}\|^2 \\
\|V_\mathcal{F}^\Phi - V_\mathcal{F}\|^2 &\leq \frac{1}{(1-\kappa)^2} \|M_\Phi V_\mathcal{F} - V_\mathcal{F}\|^2.
\end{aligned}
$$

Here, $\kappa \leq \beta$ is the contraction rate defined by the composite operator $M_\Phi \mathcal{F}$. Note that the third term follows from the second term above by Pythagoras' theorem, since $V_\mathcal{F}^\Phi - M_\Phi V_\mathcal{F}$ lies in the subspace spanned by the column space of Φ and $M_\Phi V_\mathcal{F} - V_\mathcal{F}$ lies orthogonal to this subspace. Consequently, the "distance" between the true value function V^π and the approximation $\hat{V}^\pi$ can be bounded in terms of the distance between V^π and its projection onto the space spanned by the basis functions [121]:

$$\|\hat{V}^\pi - V^\pi\|_\pi \leq \frac{1}{\sqrt{1-\kappa^2}} \|M_\Phi^\pi V^\pi - V^\pi\|_\pi .$$

The problem of value function approximation in control learning is significantly more difficult, in that it involves finding an approximate fixed point of an initially unknown operator. One standard algorithm for control learning is *approximate policy iteration* [14], which interleaves an *approximate policy evaluation* step of finding an approximation of the value function $\hat{V}^{\pi_k}$ associated with a given policy π_k at stage k, with a *policy improvement* step of finding the greedy policy associated with $\hat{V}^{\pi_k}$. Here, there are two additional sources of error introduced by approximating the exact value function, and approximating the policy. One way to eliminate one source of error is to do a least-squares approximation of the action-value function, and avoid representing the policy directly, which we turn to describe next.

6.2.2   Least-Squares Approximation of Action-Value Functions

In this section, we briefly describe least-squares methods for approximating action-value functions. In particular, we focus on an approximate policy iteration method called LSPI [67], which constructs a least-squares estimate of the true action-value function $Q^\pi(s, a)$ for a policy π using a set of (hand-coded) basis functions $\phi(s, a)$. The true action-value function $Q^\pi(s, a)$ is a vector in a high dimensional space $\mathbb{R}^{|S| \times |A|}$, and using the basis functions amounts to reducing the dimension to $\mathbb{R}^k$ where $k \ll |S| \times |A|$. The approximated action value is thus

$$\hat{Q}^\pi(s, a; w) = \sum_{j=1}^{k} \phi_j(s, a) w_j,$$

where the w_j are weights or parameters that can be determined using a least-squares method. Let Q^π be a real (column) vector $\in \mathbb{R}^{|S| \times |A|}$. $\phi(s, a)$ is a real vector of size k where each entry corresponds to the basis function $\phi_j(s, a)$ evaluated at the state action pair (s, a). The approximate action-value function can be written as $\hat{Q}^\pi = \Phi w^\pi$, where w^π is a real column vector of length k and Φ is a real matrix with $|S| \times |A|$ rows and k columns. Each row of Φ specifies all the basis functions for a particular state action pair (s, a), and each column represents the value of a particular basis function over all state action pairs. The least-squares fixed-point approximation finds a set of weights w^π under which the projection of the backed up approximated Q-function $T^\pi \hat{Q}^\pi$ onto the space spanned by the columns of Φ is a fixed point, namely

$$\hat{Q}^\pi = \Phi (\Phi^T D_{\rho^\pi} \Phi)^{-1} \Phi^T D_{\rho^\pi} (T^\pi \hat{Q}^\pi),$$

where T^π is the Bellman "backup" operator as before, and D_{ρ^π} is a diagonal matrix whose entries reflect varying "costs" for making approximation errors on state-action (s, a) pairs as a result of the nonuniform distribution $\rho^\pi(s, a)$ of visitation frequencies. It can be shown that the resulting solution can be written in a *weighted* least-squares form as $Aw^\pi = b$, where the A matrix is given by

$$A = \left(\Phi^T D_{\rho^\pi} (\Phi - \gamma P^\pi \Phi) \right),$$

and the b column vector is given by

$$b = \Phi^T D_{\rho^\pi} R.$$

A and *b* can be estimated from a database of transitions collected from some source, e.g. a random walk. The *A* matrix and *b* vector can be estimated as the sum of many rank-one matrix summations from a database of stored samples.

$$\tilde{A}^{t+1} = \tilde{A}^t + \phi(s_t, a_t) \left(\phi(s_t, a_t) - \gamma \phi(s_t', \pi(s_t')) \right)^T$$
$$\tilde{b}^{t+1} = \tilde{b}^t + \phi(s_t, a_t) r_t,$$

where (s_t, a_t, r_t, s_t') is the *t*th sample of experience from a trajectory generated by the agent (using some random or guided policy). Once the matrix *A* and vector *b* have been constructed, the system of equations $Aw^\pi = b$ can be solved for the weight vector w^π either by taking the inverse of *A* (if it is of full rank) or by taking its pseudo-inverse (if *A* is rank-deficient). The greedy policy associated with $\hat{Q}^\pi(s, a)$ is then defined as $\hat{\pi}(s) = \text{argmax}_a \hat{Q}^\pi(s, a)$, where we have $\hat{Q}^\pi = \Phi w^\pi$. The process is then repeated, until convergence (e.g., when the $\mathbb{L}^2$-normed difference between two successive weight vectors falls below a predefined threshold ϵ). Note that in successive iterations, the *A* matrix will be different since the policy π has changed. Approximation methods such as LSPI require knowing the basis functions $\phi(s, a)$ a priori. We now turn to describing a method for automatically computing these basis functions from sample trajectories.

6.3   REPRESENTATION POLICY ITERATION

This section summarizes a general framework called *Representation Policy Iteration* (RPI) for learning representation and control for solving MDPs [75]. Figure 6.1 illustrates the overall block diagram. RPI is decomposed into three components: sample collection, basis construction, and policy learning. Sample collection requires a task specification, which comprises a domain simulator (or alternatively a physically embodied agent like a robot), and an initial policy. In the simplest case, the initial policy can be a random walk, although it can also reflect a more informative hand-coded policy. The second phase involves constructing the bases from the collected samples using a diffusion model, such as an undirected (or directed) graph. This process involves finding the eigenvectors of a symmetrized graph operator such as the graph Laplacian. The final phase involves estimating the "best" policy representable in the span of the basis functions constructed (we are primarily restricting our attention to linear architectures, where the value function is a weighted linear combination of the bases). The entire process can then be iterated.

Figure 6.2 specifies a more detailed algorithmic view of the overall framework. In the sample-collection phase, an initial random walk (perhaps guided by an informed policy) is carried out to obtain samples of the underlying manifold on the state space. The number of samples needed is an empirical question. Given this set of samples, in the representation learning phase, an undirected (or directed) graph is constructed in one of the several ways:

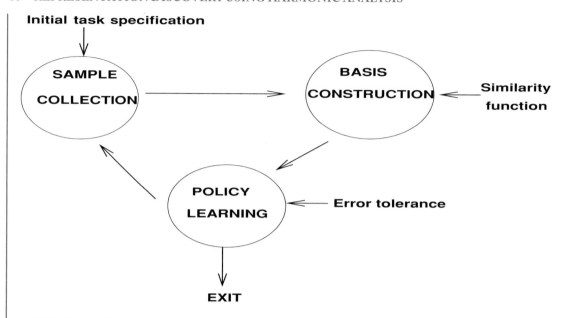

FIGURE 6.1: Flowchart of an unified approach to learning representation and behavior.

two states can be connected by a unit cost edge if they represent temporally successive states; alternatively, a local distance measure such as k-nearest neighbor can be used to connect states, which is particularly useful in the experiments on continuous domains reported below. From the graph, proto-value functions are computed using one of the graph operators, for example the combinatorial or normalized Laplacian. The smoothest eigenvectors of the graph Laplacian (that is, associated with the smallest eigenvalues) are used to form the suite of proto-value functions. The number of proto-value functions needed is a model selection question, which will be empirically investigated in the experiments described later. The encoding $\phi(s) : S \rightarrow \mathbb{R}^k$ of a state is computed as the value of the k proto-value functions on that state. To compute a state action encoding, a number of alternative strategies can be followed: the figure shows the most straightforward method of simply replicating the length of the state encoding by the number of actions and setting all the vector components to 0 except those associated with the current action.

6.3.1 Sample Run of RPI on the Two-Room Environment

The result of running the algorithm is shown in Figure 6.3, which was obtained using the following specific parameter choices.

- The state space of the two-room MDP consists of 100 states, of which 43 states are inaccessible since they represent interior and exterior walls. The remaining 57 states

RPI $(\pi_m, T, N, \epsilon, k, \mathcal{O}, \mu, \mathcal{D})$:

// π_m: Policy at the beginning of trial m
// T: Number of initial random walk trials
// N: Maximum length of each trial
// ϵ : Convergence condition for policy iteration
// k: Number of proto-value basis functions to use
// $\mathcal{O}$: Type of graph operator used
// μ: Parameter for basis adaptation
// $\mathcal{D}$: Initial set of samples

Sample-Collection Phase

- **Off-policy or on-policy sampling:** Collect a data set of samples $\mathcal{D}_m = \{(s_i, a_i, s_{i+1}, r_i), \ldots\}$ by either randomly choosing actions (off-policy) or using the supplied initial policy (on-policy) for a set of T trials, each of maximum N steps.
- **(Optional) Subsampling step:** Form a subset of samples $\mathcal{D}_s \subseteq \mathcal{D}$ by some subsampling method such as random subsampling or trajectory subsampling.

Representation Learning Phase

- Build a diffusion model from the data in $\mathcal{D}_s$. In the simplest case of discrete MDPs, construct an undirected weighted graph G from $\mathcal{D}$ by connecting state i to state j if the pair (i, j) form temporally successive states $\in S$. Compute the operator $\mathcal{O}$ on graph G, for example the normalized Laplacian $\mathcal{L} = D^{-\frac{1}{2}}(D - W)D^{-\frac{1}{2}}$.
- Compute the k smoothest eigenvectors of $\mathcal{O}$ on the graph G. Collect them as columns of the basis function matrix Φ, a $|S| \times k$ matrix.

Control Learning Phase

- Using a standard parameter estimation method (e.g. Q-learning or LSPI), find an ϵ-optimal policy π that maximizes the action-value function $Q^\pi = \Phi w^\pi$ within the linear span of the bases Φ using the training data in $\mathcal{D}$.
- **Optional:** Set the initial policy π_{m+1} to π and call RPI$(\pi_{m+1}, T, N, \epsilon, k, \mathcal{O}, \mu, \mathcal{D})$.

FIGURE 6.2: This figure shows a generic algorithm for representation discovery and control learning.

are divided into 1 doorway state and 56 interior room states. The agent is rewarded by +100 for reaching state 89, which is the last accessible state in the bottom right-hand corner of room 2, and its immediate neighbors. In the 3D value function plots shown in Figure 6.3, the axes are reversed to make it easier to visualize the value function plot, making state 89 appear in the top-left (diagonally distant) corner.

- 3463 samples were collected using off-policy sampling from a random walk of 50 episodes, each of length 100 (or terminating early when the goal state was reached).

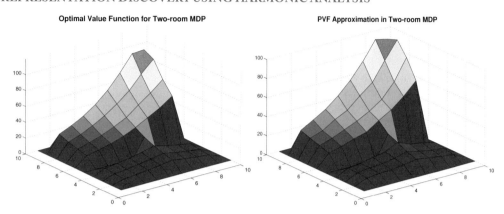

FIGURE 6.3: Optimal value function for a two-room MDP, and the approximation produced by the RPI algorithm using 20 proto-value functions, computed as the eigenvectors of the normalized graph Laplacian on the adjacency graph. The nonlinearity represented by the walls is clearly captured.

Four actions (compass direction movements) were possible from each state. Action were stochastic. If a movement was possible, it succeeded with probability 0.9. Otherwise, the agent remained in the same state. When the agent reaches state 89, or its immediate neighbors, it receives a reward of 100, and is randomly reset to an accessible interior state.

- An undirected graph was constructed from the sample transitions, where the weight matrix W is simply the adjacency $(0, 1)$ matrix. The graph operator used was the normalized Laplacian $\mathcal{L} = D^{-\frac{1}{2}} L D^{-\frac{1}{2}}$.

- 20 eigenvectors corresponding to the smallest eigenvalues of $\mathcal{L}$ (duplicated four times, one set for each action) are chosen as the columns of the state action basis matrix Φ.

- The parameter estimation method used was least-squares policy iteration (LSPI), with $\gamma = 0.8$. LSPI was described in Section 6.2.2.

- The optimal value function using unit vector bases and the approximation produced by 20 PVFs are compared in Figure 6.3.

6.3.2   Comparison with Hand-coded Parametric Bases

In this section, we compare the effectiveness of PVFs with parametric bases using small discrete MDPs, such as the two-room discrete MDP used above, before proceeding to investigate how to scale the framework to larger discrete and continuous MDPs.

Number of Basis Functions.   Figure 6.4 evaluates the learned policy by measuring the number of steps to reach the goal, as a function of the number of training episodes, and as the number

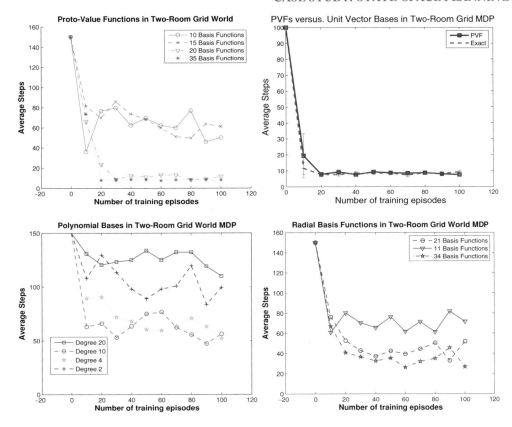

FIGURE 6.4: This experiment contrasts the performance of Laplacian PVFs (top left) with unit vector bases (top right), hand-coded polynomial basis functions (bottom left), and radial basis functions (bottom right) on a 100 state two-room discrete MDP. Results are averaged over 10 runs. The performance of PVFs (with 25 bases) closely matches that of unit vector bases, and is considerably better than both polynomials and RBFs on this task.

of basis functions is varied (ranging from 10 to 35 for each of the four actions). The results are averaged over 10 independent runs, where each run consisted of a set of training episodes of a maximum length of 100 steps, where each episode was terminated if the agent reached the absorbing goal state. Around 20 basis functions (per action) were sufficient to get close to optimal behavior, and increasing the number of bases to 35 produced a marginal improvement. The variance across runs is fairly small for 20 and 35 bases, but relatively large for smaller numbers of bases (not shown for clarity). Figure 6.4 also compares the performance of PVFs with unit vector bases (table lookup), showing that PVFs with 25 bases closely tracks the performance of unit vector bases on this task.

Comparison with Parametric Bases.   One important consideration in evaluating PVFs is how they compare with standard parametric bases, such as radial basis functions and polynomials. Figure 6.4 evaluates the effectiveness of polynomial bases and radial basis functions in the two-room MDP. In polynomial bases, a state i is mapped to the vector $\phi(i) = (1, i, i^2, \ldots i^{k-1})$ for k basis functions—this architecture was studied in [63, 67]. In RBFs, a state i is mapped to $\phi_j(i) = \exp^{-\frac{(i-j)^2}{2\sigma^2}}$, where j is the center of the RBF basis function. In the experiments shown, the basis centers were placed equidistantly from each other along the 100 states. The results show that both parametric bases under these conditions performed worse than PVFs in this task.

6.4   SCALING PROTO-VALUE FUNCTIONS: PRODUCT SPACES

In this section, we build on the general framework for scaling representation discovery to large *factored* spaces described in Chapter 5, where we exploit the spectral properties of the graph Laplacian in constructing embeddings that are highly regular for structured graphs (see Figure 5.2). In particular, as we saw previously, the eigenspace of the Kronecker sum of two graphs is the Kronecker product of the eigenvectors of each component graph.

6.4.1   Factored Representation Policy Iteration for Structured Domains

We derive the update rule for a factored form of RPI (and LSPI) for structured domains when the basis functions can be represented as Kronecker products of elementary basis functions on simpler state spaces. Basis functions are *column* eigenvectors of the diagonalized representation of a graph operator, whereas embeddings $\phi(s)$ are *row* vectors representing the first k basis functions evaluated on state s. By exploiting the property that $(A \otimes B)^T = A^T \otimes B^T$, it follows that embeddings for structured domains can be computed as the Kronecker products of embeddings for the constituent state components. As a concrete example, a grid world domain of size $m \times n$ can be represented as a graph $G = G_m \oplus G_n$, where G_m and G_n are *path graphs* of size m and n, respectively. The basis functions for the entire grid world can be written as the Kronecker product $\phi(s) = \phi_m(s^r) \otimes \phi_n(s^c)$, where $\phi_m(s^r)$ is the basis (eigen)vector derived from a path graph of size m (in particular, the row s^r corresponding to state s in the grid world), and $\phi_n(s^c)$ is the basis (eigen)vector derived from a path graph of size n (in particular, the column s^c corresponding to state s in the grid world).

Extending this idea to state action pairs, the basis function $\phi(s, a)$ can be written as $e_I(a) \otimes \phi(s)$, where $e_I(a)$ is the unit vector corresponding to the index of action a (e.g., action a_1 corresponds to $e_1 = [1, 0, \ldots]^T$). Actually, the full Kronecker product is not necessary if only a relatively small number of basis functions are needed. For example, if 50 basis functions are to be used in a $10 \times 10 \times 10$ hypercube, the full state embedding is a vector of size 1000, but only the first 50 terms need to be computed. Such savings imply proto-value functions can be efficiently computed even in very large structured domains. For a factored state space $s = (s^1, \ldots, s^m)$, we use the notation s^i to denote the value of the ith component. We can restate the update rules for factored RPI and LSPI as follows:

$$\tilde{A}^{t+1} = \tilde{A}^t + \phi(s_t, a_t)\left(\phi(s_t, a_t) - \gamma\phi(s_t', \pi(s_t'))\right)^T$$
$$= \tilde{A}^t + e_{I(a_t)} \otimes \prod_\otimes \phi_i(s_t^i)$$

$$\times \left(e_{I(a_t)} \prod_\otimes \phi_i(s_t^i) - \gamma e_{I(\pi(s_t'))} \otimes \prod_\otimes \phi_i(s_t'^i)\right)^T .$$

The corresponding update equation for the reward component is

$$\tilde{b}^{t+1} = \tilde{b}^t + \phi(s_t, a_t)r_t = \tilde{b}^t + r_t e_{I(a_t)} \otimes \prod_\otimes \phi_i(s_t^i).$$

6.4.2   Experimental Results

We now present a detailed study using a much larger factored multiagent domain called the "Blockers" task, which was studied in [103]. This task, illustrated in Figure 6.5, is a cooperative multiagent problem where a group of agents try to reach the top row of a grid, but are prevented in doing so by "blocker" agents who move horizontally on the top row. If any agent reaches the top row, the entire team is rewarded by +1; otherwise, each agent receives a negative reward of −1 on each step. The agents always start randomly placed on the bottom row of the grid, and the blockers are randomly placed on the top row. The blockers remain restricted to the top row, executing a fixed strategy. The overall state space is the Cartesian product of the location of each agent. These experiments on the blocker domain include more difficult versions of the

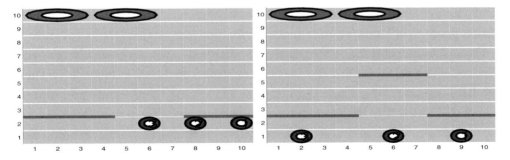

FIGURE 6.5: Two versions of the blocker domain are shown, each generating a state space of $> 10^6$ states. Interior walls shown create an "irregular" factored MDP whose overall topology can be viewed as a "perturbed" variant of a pure product space.

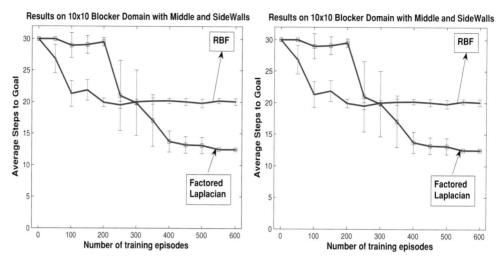

FIGURE 6.6: Comparison of factored (Laplacian) PVF basis functions with hand-coded radial basis functions (RBF) on a 10×10 "wrap-around" grid with three agents and two blockers of $>10^6$ states. Both approaches were tested using 100 basis functions. The plots show the performance of PVFs against RBFs on the two blocker domains in Figure 6.5.

task not studied in [103] specifically designed to test the scalability of the Kronecker product bases to "irregular" grids whose topology deviates from a pure hypercube or toroid. In the first variant, shown on the left in Figure 6.5, horizontal interior walls extend out from the left and right side walls between the second and third row. In the second variant, an additional interior wall is added in the middle as shown on the right.[1]

The basis functions for the overall Blocker state space were computed as Kronecker products of the basis functions over each agent's state space. Each agent's state space was modeled as a grid or a cylinder (for the "wrap-around" case). Since the presence of interior walls obviously violates the pure product of cylinders or grids topology, each individual agent's state space was learned from a random walk. The overall basis functions were then constructed as Kronecker products of Laplacian basis functions for each learned (irregular) state grid.

Figure 6.6 compares the performance of the factored Laplacian bases with a set of radial basis functions (RBFs) for the first Blocker domain (shown on the left in Figure 6.5). The width of each RBF was set at $\frac{2|S_a|}{k}$, where $|S_a|$ is the size of each individual agent's grid, and k is the number of RBFs used. The RBF centers were uniformly spaced. The results shown are averages over ten learning runs. On each run, the learned policy is measured every 25 training episodes. Each episode begins with a random walk of a maximum of 70 steps (terminating earlier if the

---

[1]In the Blocker domain, the interior walls are modeled as having "zero width", and hence all 100 states in each grid remain accessible, unlike the two-room environment.

top row was reached). After every 25 such episodes, RPI is run on all the samples collected thus far. The learned policy is then tested over 500 test episodes. The graphs plot the average number of steps to reach the goal. The experiments were conducted on both "normal" grids (not shown) and "wrap-around" cylindrical grids. The results show that RBFs converge faster, but learn a worse policy. The factored Laplacian bases converge slower than RBFs, but learn a substantially better policy. Figure 6.6 also shows results for the second Blocker domain (shown on the right in Figure 6.5 with both side and interior middle walls), comparing 100 factored Laplacian bases with a similar number of RBFs. The results show a significant improvement in the performance of the factored Laplacian bases over RBFs.

In terms of both space and time, the factored approach greatly reduces the computational complexity of finding and storing the Laplacian bases. A worst-case estimate of the size of the full Laplacian matrix is $O(|S|^2)$. Diagonalizing a $|S| \times |S|$ symmetric matrix and finding k eigenvectors requires time $O(k|S|^2)$ and $O(k|S|)$ space. Instantiating these general estimates for the Blocker domain, let n refer to the number of rows and columns in each agent's state space ($n = 10$ in these experiments), and k refer to the number of basis functions ($k = 100$ in these experiments). Then, the size of the state space is $|S| = (n^2)^3$, implying that the non-factored approach requires $O(k(n^2)^3)$ space and $O(k(n^6)^2)$ time, whereas the factored approach requires $O(kn^2)$ space and $O(k(n^2)^2)$ time. Note these are the worse-case estimates. The Laplacian matrix is in fact highly sparse in the Blocker domain, requiring far less than $O(|S|^2)$ space to be stored. In fact, even in such a deterministic MDP where the Laplacian matrix can be stored in $O(|S|)$ space, the non-factored approach will still take $O(kn^3)$ space and $O(kn^6)$ time, whereas the factored approach takes $O(kn)$ space and $O(kn^2)$ time.

6.5   RPI IN CONTINUOUS DOMAINS

In this section, we present an experimental analysis of fully interleaved representation discovery and policy learning on continuous MDPs [79]. By "fully interleaved", we mean that the overall learning run is divided into a set of discrete episodes of sample collection, basis construction, and policy learning. At the end of each episode, a set of additional samples is collected using either a random walk (off-policy) or the currently best-performing policy (on-policy), and then basis functions are then recomputed and a new policy is learned.

6.5.1   Three Control Tasks

We explore the effectiveness and stability of proto-value functions in three continuous domains—the Acrobot task, the inverted pendulum task, and the mountain car task—that have long been viewed as benchmarks in the field [111]. These three domains are now described in more detail.

The Inverted Pendulum.   The inverted pendulum problem requires balancing a pendulum of unknown mass and length by applying force to the cart to which the pendulum is attached. We used the implementation described in [67]. The state space is defined by two variables: θ, the vertical angle of the pendulum, and $\dot{\theta}$, the angular velocity of the pendulum. The three actions are applying a force of -50, 0, or 50 N. Uniform noise from -10 and 10 is added to the chosen action. State transitions are defined by the nonlinear dynamics of the system, and depend upon the current state and the noisy control signal, u.

$$\ddot{\theta} = \frac{g\sin(\theta) - \alpha ml\dot{\theta}^2 \sin(2\theta)/2 - \alpha\cos(\theta)u}{4l/3 - \alpha ml \cos^2(\theta)}, \tag{6.5}$$

where g is the gravity, 9.8 m/s^2, m is the mass of the pendulum, 2.0 kg, M is the mass of the cart, 8.0 kg, l is the length of the pendulum, 0.5 m, and $\alpha = 1/(m + M)$. The simulation time step is set to 0.1 s. The agent is given a reward of 0 as long as the absolute value of the angle of the pendulum does not exceed $\pi/2$. If the angle is greater than this value the episode ends with a reward of -1. The discount factor was set to 0.95. The maximum number of episodes the pendulum was allowed to balance was fixed at 3000 steps. Each learned policy was evaluated ten times.

Mountain Car.   The goal of the *mountain car* task is to get a simulated car to the top of a hill as quickly as possible [111]. The car does not have enough power to get there immediately, and so must oscillate on the hill to build up the necessary momentum. This is a minimum time problem, and thus the reward is -1 per step. The state space includes the position and velocity of the car. There are three actions: full throttle forward ($+1$), full throttle reverse (-1), and zero throttle (0). Its position, x_t and velocity $\dot{x}_t$, are updated by

$$x_{t+1} = \text{bound}[x_t + \dot{x}_{t+1}], \tag{6.6}$$

$$\dot{x}_{t+1} = \text{bound}[\dot{x}_t + 0.001a_t + -0.0025, \cos(3x_t)], \tag{6.7}$$

where the bound operation enforces $-1.2 \leq x_{t+1} \leq 0.6$ and $-0.07 \leq \dot{x}_{t+1} \leq 0.07$. The episode ends when the car successfully reaches the top of the mountain, defined as the position $x_t \geq 0.5$. In the experiments we allow a maximum of 500 steps, after which the task is terminated without success. The discount factor was set to 0.99.

The Acrobot Task.   The Acrobot task [111] is a two-link under-actuated robot that is an idealized model of a gymnast swinging on a highbar. The only action available is a torque

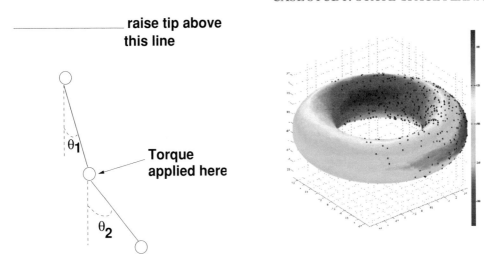

FIGURE 6.7: The state space of the Acrobot (shown on the left) exhibits rotational symmetries. The figure on the right plots its projection onto the subspace of $\mathbb{R}^2$ spanned by the two joint angles θ_1 and θ_2, which can be visualized as a torus. The angular velocities $\dot{\theta}_1$ and $\dot{\theta}_2$ were set to 0 for this plot. The points shown on the torus are subsampled states from a random walk. The colors indicate the value function, with red (darker) regions representing states with higher values.

on the second joint, discretized to one of the three values (positive, negative, and none). The reward is -1 for all transitions leading up to the goal state. The detailed equations of motion are given in [111]. The state space for the Acrobot is four-dimensional. Each state is a four-tuple represented by $(\theta_1, \dot{\theta}_1, \theta_2, \dot{\theta}_2)$. θ_1 and θ_2 represent the angle of the first and second links to the vertical, respectively, and are naturally in the range $(0, 2\pi)$. $\dot{\theta}_1$ and $\dot{\theta}_2$ represent the angular velocities of the two links. Note that angles near 0 are actually very close to angles near 2π due to the rotational symmetry in the state space.

Figure 6.7 plots the Acrobot state space projected onto the subspace spanned by the two joint angles θ_1 and θ_2. This subspace is actually a torus. To approximate computing distances on the torus, the original states were projected upwards to a higher dimensional state space $\subset \mathbb{R}^6$ by mapping each angle θ_i to $(\sin(\theta_i), \cos(\theta_i))$. Thus, the overall state space is now $(\sin(\theta_1), \cos(\theta_1), \dot{\theta}_1, \sin(\theta_2), \cos(\theta_2), \dot{\theta}_2)$. The motivation for this remapping is that now Euclidean distances in this augmented space better approximate local distances on the torus. In fact, ignoring the wrap-around nature of the Acrobot state space by simply using a local Euclidean distance metric on the four-dimensional state space results in significantly poorer performance. This example illustrates how overall global knowledge of the state space, just like in the Blockers domain, is valuable in designing a better local distance function for learning PVFs.

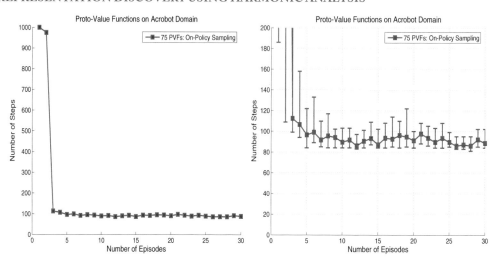

FIGURE 6.8: The performance of PVFs with on-policy sampling in the Acrobot task. The plot on the left shows the median average number of steps to goal averaged over 30 runs. The plot on the right shows the variance, after scaling the y-axis to magnify the plot.

6.5.2   RPI with On-Policy Sampling

The performance of PVFs can be improved using a modified form of on-policy sampling in Step 1 of the sample-collection phase in the RPI algorithm. Specifically, it requires keeping track of the best-performing policy (in terms of the overall performance measure of the number of steps). If the policy learned in the current round of RPI improved on the best-performing policy thus far, samples were collected in the next iteration of RPI using the newly learned policy (which was then viewed as the best performing policy in subsequent runs). Otherwise, samples were collected using an off-policy random walk. We also found that using shorter episodes of sample collection in between rounds of representation construction and policy estimation also produced better results. Figure 6.8 shows the results of these two modifications in the Acrobot domain, where convergence is fairly rapid.

6.5.3   Comparing PVFs with RBFs on Continuous MDPs

In this section, we compare the performance of PVFs with radial basis functions (RBFs), which are a popular choice of basis functions for both discrete and continuous MDPs. We restrict the comparison of PVFs and RBFs in this section to the inverted pendulum domain. To choose a suitable set of parameters for RBFs, we manually tuned the kernel widths, to find a good choice for these parameters. The comparison shown in Figure 6.9 shows that PVFs are significantly quicker to converge, by almost a factor of two in the inverted pendulum domain. Asymptotically, both approaches converge to the same result. PVFs take 20 trials to converge,

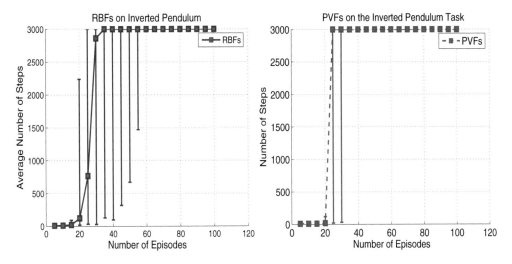

FIGURE 6.9: This plot shows that PVFs (right) have significantly less variance compared to RBFs (left) in the inverted pendulum task. Both plots show median-averaged number of steps the pole was balanced over 100 learning runs.

but RBFs take roughly twice as long. Figure 6.9 plots the variance across 100 learning runs for both PVFs and RBFs, showing that PVFs not only converge faster, but also have significantly less variance.

Figure 6.10 shows the variances over 30 runs for both PVFs and RBFs in the mountain car domain. As in the inverted pendulum, we note that PVFs clearly converge more quickly to a more stable performance than RBFs, although the differences are not as dramatic as in the inverted pendulum domain.

6.5.4   Policy and Reward-Sensitive PVFs

In the experiments presented above, basis functions are constructed without taking rewards into account. This restriction is not intrinsic to the approach, and reward or policy information when available can easily be incorporated into the construction of bases. There are also many alternative settings to the one we have explored. One approach studied in [95] assumes that the reward function R^π and policy transition matrix P^π are known, and combines low-order Laplacian eigenvector bases with *Krylov bases*. This approach is restricted to *policy evaluation*, which consists of solving an equation in the well-studied form $Ax = b$. Krylov bases are used extensively in the solution of such linear systems of equations [47]. The Krylov space is defined as the space spanned by the vectors:

$$K_m(A, b) = \langle b \ Ab \ A^2b \ \dots A^{m-1}b \rangle.$$

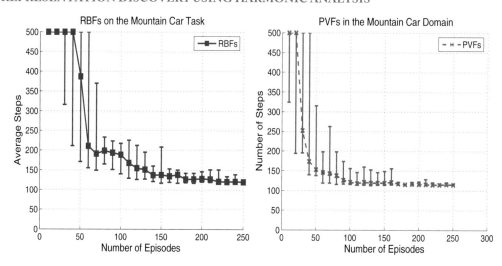

FIGURE 6.10: Left: The variance in the performance of a linear parametric RBF architecture is analyzed over 30 learning runs in the mountain car domain. Right: Variance across 30 runs for PVFs in the mountain car task.

Krylov bases have also been used to compress the *belief* space of a partially-observable Markov decision process (POMDP) [97]. Finally, Keller et al. [62] and Parr et al. [93] explore *Bellman error basis functions* (BEBFs) by explicitly using the error in approximating the value function using the *Bellman residual* $T^\pi(V) - V$. At every step of policy evaluation, the next basis added is proportional to the estimated Bellman residual. Since T^π is not known in control learning, the Bellman residual is approximated by the sample error $r + \gamma V(s') - V(s)$. BEBFs, like Krylov bases, can be more effective than PVFs since they are tailored to the Krylov subspace spanned by powers of the transition matrix P^π and the reward function. However, this dependence also means they need to be recomputed for each specific policy and reward function, with no savings accruing from sharing across related tasks on the same state space.

6.5.5   Extensions of Proto-Value Functions

For simplicity, thus far, we have explored the construction of basis functions using Laplacian eigenvectors in the simplest setting of undirected graphs. Johns et al. [55] show that this approach readily generalizes to directed graphs, where the directed Laplacian is used to construct directional basis functions. Also, approximation of action-value functions requires constructing state-action basis functions. The approach above assumed that state-dependent basis functions $\phi(s)$ are copied $|A|$ times to construct state-action basis functions. A more elegant approach is to construct state-action bases by diagonalizing diffusion operators on graphs whose vertices are state-action pairs [92]. Finally, the Kronecker sum decomposition method described in

Section 6.4.1 assumed that the global structure of the state space is known. One approach to automatically constructing factored bases by discovering Kronecker product factorizations of weight matrices is explored in [56].

6.6   MULTISCALE BASIS CONSTRUCTION FOR MARKOV DECISION PROCESSES

In this section, we describe one specific application of the general diffusion wavelet framework introduced in Chapter 4 to Markov decision processes. In particular, we show how diffusion wavelets provide a novel algorithm for policy evaluation [72], a critical step in the solution of large Markov decision processes (MDPs), typically requiring $O(|S|^3)$ to directly solve the Bellman system of $|S|$ linear equations (where $|S|$ is the number of samples of the underlying state space). For a fixed policy π, this framework efficiently constructs a multiscale decomposition of the random walk P^π associated with the policy π. This enables efficiently computing medium and long term state distributions, approximation of value functions, and the *direct* computation of the potential operator $(I - \gamma P^\pi)^{-1}$ needed to solve Bellman's equation. Even a preliminary non-optimized version of the solver is shown to be competitive with highly optimized iterative techniques. In particular, for MDPs where the transition matrix is "diffusion-like", the asymptotic complexity of building a diffusion wavelet tree is $O(|S| \log^2 |S|)$.

6.6.1   Preliminaries

Bellman's equation usually involves the solution of a sparse linear system of size $|S|$, where S is the state space. A classical direct solution of the system is infeasible for large problem sizes, since it requires $\mathcal{O}(|S|^3)$ steps. One common technique is to use an iterative method, such as value iteration, which has the worst case complexity $\mathcal{O}(|S|^2)$ for sparse transition matrices, and $\mathcal{O}(|S| \log |S|)$ when the problem is well-conditioned and only low-precision is required. The approach in this section is fundamentally different, and yields a *direct* solution in time $\mathcal{O}(|S| \log^2 |S|)$. It consists of two parts:

(i) a *pre-computation* step that depends on the structure of the state space and on the policy. The result of this step is a multiscale hierarchical decomposition of the value function space over the state space, and a multiscale compression of powers of the transition matrix (random walk operator) over the state space. This computation, for classes of problems of interest in applications, has complexity $\mathcal{O}(|S| \log^2 |S|)$.

(ii) an *inversion* step that uses the multiscale structure built in the "pre-computation" step to efficiently compute the solution of Bellman's equations for a given reward function. This phase of the computation has complexity $\mathcal{O}(|S| \log |S|)$ for many problems of practical importance where the transition matrix is *diffusion*-like (defined precisely below). The constants in front of this asymptotic complexity are much smaller than those in the pre-computation step.

We will define the class of problems for which the complexity of the diffusion wavelet method is linear up to logarithmic factors. Qualitatively, this class includes the case of state spaces that can be represented by a finite undirected weighted graph, with all the vertices of "small" degree in which transitions are allowed only among neighboring points, and the spectrum of the transition matrix decays fast enough. The direct method we present offers several advantages.

(i) The multiscale construction allows efficient approximation of reward and value functions, which is an important task *per se* [67, 75].

(ii) It is well known that the number of iterations necessary for an iterative method to converge can be very large, depending on the condition number of the problem (which in general depends on the number of samples), and on the precision required. Increasing precision in the direct inversion technique can be done more efficiently. In this context, even a non-optimized implementation of the diffusion wavelet approach outperforms standard iterative solvers.

(iii) When the state space and the policy are fixed, and many value functions corresponding to different rewards (tasks) need to be computed, iteration schemes do not take advantage of the common structure between the problems. In this case, the number of iterations for finding each solution is multiplied by the number of solutions sought. The diffusion wavelet direct inversion technique efficiently encodes the common structure of the state space in the pre-computation step, and then takes advantage of this in the solution of multiple problems.

A key advantage of the proposed approach is that direct inversion reveals interesting structure in the underlying problem. The multi-resolution construction has interesting connections to methods for approximately solving hierarchical Markov decision processes [8].

6.6.2 Direct Solution of Bellman's Equation

This section describes a multiscale approach that allows a direct solution of Bellman's equation. The starting point are the identities

$$V^\pi = (I - \gamma P^\pi)^{-1} R = \sum_{k \geq 0} (\gamma \Pi^{-\frac{1}{2}} T^\pi \Pi^{\frac{1}{2}})^k R$$
$$= \prod_{k \geq 0} (I + \gamma^{2^k} \Pi^{-\frac{1}{2}} (T^\pi)^{2^k} \Pi^{\frac{1}{2}}) R, \tag{6.8}$$

where $P^\pi = \Pi^{-\frac{1}{2}} T^\pi \Pi^{\frac{1}{2}}$, Π is the matrix whose diagonal is the asymptotic distribution of P, and R is the reward vector. The first identity follows by the definition of V^π, and the second is the usual Neumann series expansion for the inverse. The last identity is called the *Schultz formula*, which is true because each term of the Neumann series appears once and only once in the product (reordering the terms of the summation is allowed because both the sum and the product are absolutely convergent). The formulas hold for $\gamma \leq 1$ and R^π not in the kernel of $(I - \gamma P^\pi)$. The sums and products involved are of course finite once the precision is fixed.

A key component underlying the diffusion wavelet approach is the compression of the (quasi-)dyadic powers of the operator T^π. In particular $(T^\pi)^{2^j-1} f$, for any function f, is equal to the product $R_j R_{j-1} \cdots \cdots R_0 f$, where R_i represents the operator $(T^\pi)^{2^i}$ on the basis Φ_i in the domain and Φ_{i+1} in the range, and hence the product above is $T^{1+2+2^2+\cdots+2^{j-1}} f = T^{2^j-1} f$, represented on Φ_{j+1}, i.e. "in compressed form". The matrices $[\Phi_{j+1}]^*_{\Phi_j}$ "un-pack" this representation back onto the basis Φ_0. To obtain $T^{2^j} f$ we only need to apply T once more. In this way the computation of $T^{2^j} f$ takes only $\mathcal{O}(j|S|\log|S|)$ operations, since R_j contains about $\mathcal{O}(|S|\log|S|)$ entries. This cost should be compared to that of computing directly the matrix T^{2^j}, which is $\mathcal{O}(2^j|S|)$ since this matrix contains about $\mathcal{O}(2^j|S|)$ nonzero entries; this is also the cost of applying the matrix T to f 2^j times.

In iterative methods such as value iteration, up to $|S|$ iterations are necessary, and the cost is thus $\mathcal{O}(|S|^2)$. The diffusion wavelet technique has asymptotic complexity $\mathcal{O}(|S|\log^2|S|)$. In some cases far fewer than $|S|$ iterations are needed, especially when the problem is well conditioned (e.g. γ far from 1), and of low precision. Even in this case, the diffusion wavelet approach offers several advantages, as discussed above, in terms of understanding the structure of the problem, of creating useful basis functions, and is competitive in terms of speed, as we show in the experiments.

6.6.3   Experiments

In this section, we illustrate the multiscale analysis on a sample set of MDPs on a continuous two-room spatial environment, where the two rooms have an elongated shape and are connected by a corridor. The shape of the rooms and corridor is quite arbitrary, the bases are built automatically, so we do not require any special topology or shape for them (except connectedness, without the loss of generality). The agent has randomly explored the space, so S consists of $|S|$ randomly scattered points in the rooms (see Figure 6.11). The rooms could have been of arbitrary shape and dimension, as the only input to the algorithm is the set of sampled points (vertices) and the local distances between close-by points (edge weights). The weights correspond to a natural diffusion associated with the random walk in the two rooms, restricted to the states S actually explored, by setting $W(i, j) = e^{-2||x_i - x_j||^2}$. We then construct

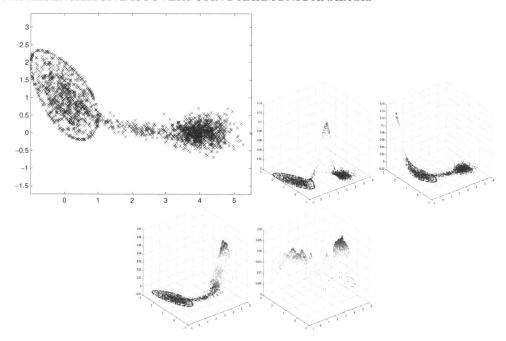

FIGURE 6.11: Top: set of samples in a continuous two-room environment. Bottom: four diffusion scaling functions built on the set, at increasing scale. Note the localization at the finest scales, and the global support at the coarsest scale.

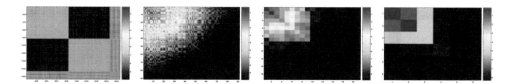

FIGURE 6.12: Compression of the powers of the symmetrized random walk T in the two-room environment. From top left to bottom right by rows: T_0, T_1, T_4, and T_6. All the matrices are represented in the $\log_{10}$ scale. T_0 is sorted to show the two-room and corridor structures (the algorithm is of course independent of the order of the points): the two large blocks represent transitions within each room, and the bottom-right block are transitions in the corridor, with bands at the bottom and at the right indicating the transitions from the corridor to the rooms. Note the decreasing size of the matrices. T_6 is very small, and essentially represents only the transition between two states (the two rooms): for time scales of order 2^6 the algorithm has automatically decided this representation is faithful enough for the precision requested.

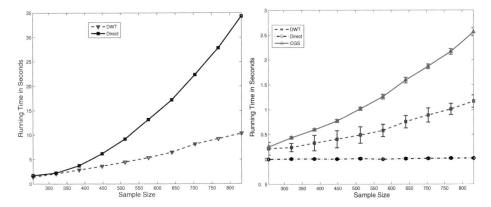

FIGURE 6.13: Left: mean and standard deviation of running time for solving a Bellman equation on a random walk in the two-room environment, as a function of the number of states explored (*x*-axis). We compared direct DWT inversion, iterative Conjugate Gradient Squared method (Matlab implementation) and direct inversion. Left: pre-processing time, comparing computation of the full inverse and construction of the diffusion wavelet tree. Right: computation time of applying the inversion scheme, comparing direct inverse, Schultz's method with diffusion wavelet transform, and symmetric conjugate gradient. Note the scale difference between the two plots.

the corresponding multiscale analysis, with precision set to 10^{-10}. Figure 6.11 and Figure 6.12 present some of the scaling functions obtained, and the compressed representation of powers of T, respectively. We then pick a random reward R on S (a vector of white Gaussian noise), and compute the corresponding value function in three ways: (i) direct computation of the matrix $I - \gamma P^\pi$, (ii) Schultz's method and diffusion wavelet transform as in (6.8), and (iii) conjugate gradient descent for symmetric matrices.

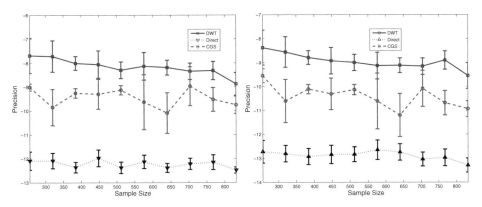

FIGURE 6.14: Precision, defined as $\log_{10}$ of the residual error $||(I - \gamma P^\pi)\tilde{V}^\pi - R||_p$, where $\tilde{V}^\pi$ is the computed solution, achieved by the different methods. The precision was set at $\varepsilon = 10^{-10}$. We show the results for $p = 2$ (left) and $p = \infty$ (right).

In this example we set $\gamma = 1$. We repeat the above for $|S| = 256$ to $|S| = 832$ in steps of 64, and for each S, for ten randomly generated rewards R. The first two methods are direct: we look at both the pre-processing time for computing, respectively, the inverse matrix and the diffusion wavelet tree (see Figure 6.13, left). We compare over several random choices of the reward vector the mean time and standard deviation for computing the corresponding value function, with all the three methods: see Figure 6.13, right. Finally, in Figure 6.14 we show the L^2- and L^∞-norms of the residual error $((I - P^\pi)\tilde{V}^\pi - R$, where $\tilde{V}^\pi$ is the estimated value function), achieved by the three methods.

6.7   BIBLIOGRAPHICAL REMARKS

A detailed overview of classical methods for solving Markov decision processes is given in [98]. Approximation methods for solving Markov decision processes using value functions are described in [14, 111]. Much of this chapter is based on [79]. Section 6.6 is based on [72, 73].

CHAPTER 7

Case Study: Computer Graphics

Representation discovery via construction of basis functions has significant commercial potential: an appropriate choice of basis can result in a highly *compressed* representation of Internet content, such as images and movies. A classic application of harmonic analysis in this regard are the well-known JPEG and JPEG-2000 standards [122], widely used by millions of consumers every day in digital cameras and on the Internet. JPEG relies on the discrete cosine transform [3], a type of Fourier analysis on 2D arrays, and JPEG-2000 relies on the wavelet transform. Both these methods, however, do not generalize to 3D objects with arbitrary topology, a problem of much current interest in applications such as computer graphics and animation.

In this chapter, we show how representation discovery using harmonic analysis can lead to the development of new approaches to the compression of 3D multimedia content. We focus on one specific problem in 3D graphics, namely *mesh compression* [112, 113]: 3D objects are specified by their graph topology and their mesh geometry specifying the location of each vertex in 3D. The space taken by these objects in the default "unit vector" basis is very large, and can be hundreds of megabytes. By generalizing Fourier and wavelet analysis to graphs, it is possible to construct object-specific basis functions that provide the customized compression of individual objects. These basis functions can be constructed during run-time, resulting in a highly sparse representation.

In this domain, we will see that the wavelet approach offers a distinct performance improvement over the Fourier approach for objects whose 3D mesh geometry is highly nonlinear, that is for mesh surfaces where the geometry changes from a relatively smooth function to a highly non-smooth function (e.g. consider the geometry of an animal like an "elephant"). We compare mesh compression using basis functions constructed using Fourier (Laplacian) eigenvectors versus using diffusion wavelets [76]. We also show one possible approach to scaling these approaches to large 3D objects, using graph partitioning, where a given 3D object is partitioned into a large number of small patches, and the basis functions are then derived on each individual patch [59].

7.1   INTRODUCTION

JPEG compression is widely used to capture, store, and distribute images [122]. Formally, JPEG uses the discrete cosine transform [3] to convert images from a "spatial" basis to a Fourier basis, where much of the image "energy" is concentrated in the low-frequency eigenvectors. However, DCT assumes a fixed 2D topology and cannot be directly applied to compress 3D objects in computer animation and graphics. Consequently, the problem of compression of 3D objects is of much interest in computer graphics [112, 113]. As we saw in Chapter 3, Fourier analysis can be easily extended to graphs, where the eigenvectors of the graph Laplacian are used as basis functions. This provides an adaptive spectral compression method, where the compression is customized to specific 3D objects by deriving basis functions from the object's known graph topology [59].

However, the compression of 3D objects is a challenging problem for harmonic analysis. 3D objects can be very large, resulting in graphs with 10^5 or more vertices, and millions of edges. Computing eigenvectors of matrices resulting from such large graphs is clearly intractable. Furthermore, even if such large matrices could be diagonalized, the resulting bases, being global in nature, are extremely large. Storing each eigenvector requires space $O(|V|)$, and seems impractical for graphs of size 10^5 or larger. The challenge of 3D compression does not stop there: typically, problems of interest in graphics also require approximating high-dimensional objects, where entire matrices stored at each vertex representing texture, lighting, etc, need to be compressed [124].

In this chapter, we explore some ways of addressing these challenges. First, since Fourier methods are based on global eigenvector representations, they do not yield a multi-resolution analysis, and poorly capture "transients" and "local discontinuities". As Figure 7.1 illustrates, these limitations have tangible consequences: it is hard to efficiently approximate piecewise smooth mesh geometries, such as the nonlinearities represented by "horns". We compare the performance of multiscale diffusion wavelet bases introduced in Chapter 4 with the Laplacian eigenvector-based approach, which was introduced in Chapter 3. To deal with the challenge of large graphs, we use a graph-partitioning method, which was suggested in Chapter 5, where the complete graph is decomposed into a set of subgraphs, and local basis functions are constructed on each subgraph.

7.2   SPECTRAL MESH COMPRESSION: FOURIER VERSUS WAVELET BASES

Figure 7.1 vividly illustrates the difference between using the Fourier approach versus the wavelet approach [76]. Laplacian eigenvectors poorly reconstruct local nonlinearities represented by the "horns" or the "nose", which are rendered with much higher fidelity by

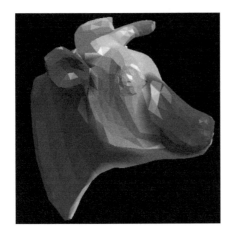

FIGURE 7.1: Spectral approximation of the mesh geometry of a 3D object using Laplacian eigenvectors (left) versus diffusion wavelet bases (right), both specifically constructed for this object.

diffusion wavelet bases. This object has 1107 vertices, which were partitioned into 10 subgraphs, and 20 basis functions were used to approximate the mesh geometry on each subgraph. Colors indicate partitions of the object on which both basis functions were computed.

Figure 7.2 illustrates some sample diffusion wavelet bases for the "cowhead" model. Here, the basis functions are shown illuminated over a darkly shaded region over the set of vertices beginning with (left) levels 4, 5, 8, and (right) 9 of the diffusion wavelet hierarchy. Note that as expected, the wavelet bases at lower levels has local support, and in fact, can be located on semantically interesting regions of the object. In addition, because basis functions are constructed at varying resolutions and spatial scales, they are more adept at approximating piece-wise linear functions, as the theoretical analysis in Chapter 4 suggested. Figure 7.3 shows two sample scaling functions from level 5, one localized in the "eye" region and the other localized in the "horn" region.

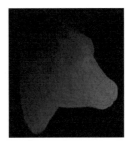

FIGURE 7.2: Scaling functions from multiple levels of the diffusion wavelet hierarchy for a sample 3D object.

FIGURE 7.3: Some sample scaling functions localized in "semantically" meaningful regions of objects.

7.3 APPROXIMATION OF MESH GEOMETRY USING OBJECT-SPECIFIC BASES

Let us define the problem of compression in 3D computer graphics more formally. Each 3D object has an "initial" representation, where the mesh geometry is specified by $3N$ floating point numbers (N is the number of vertices). To specify the topology of the object requires additionally $O(kN)$ space (since each vertex is usually connected to a small number of neighbors in polygonal meshes, where typically $k \leq 10$). The goal is to construct a basis matrix Φ of size $N \times m$, where $m \ll N$, such that the mesh geometry can be approximated to the desired level of accuracy by least-squares projection on the column space of the matrix Φ. This implies that only $3m \ll 3N$ numbers will be needed to specify the mesh geometry.

More formally, the problem of mesh compression is to approximate the 3D coordinate functions mapping each vertex to its 3D position $V \rightarrow \mathbb{R}^3$ [112, 113]. A 3D object is specified by a graph $G = (V, E, W, M)$, where the 3D mesh coordinates $M(v) \in \mathbb{R}^3$. The weight matrix W is a set of weights on each edge $e \in E$. In the experiments below, we used binary weights so that $W(i, j) = 1$ if $(i, j) \in E$. The problem is to approximate the mesh coordinates using a set of basis functions that can be computed from the weight matrix of the graph. More precisely, let v_x, v_y, v_z be the coordinates of a vertex $v \in G$. Each of these coordinate functions can be approximated by projecting them on the subspace spanned by the columns of Φ, corresponding to either a Fourier (Laplacian) or a (diffusion) wavelet basis.

7.3.1 Global Laplacian Eigenfunctions

ptFollowing the approach described in Chapter 3, global Fourier basis functions can be constructed on a graph $G = (V, E, W)$ by diagonalizing the combinatorial graph Laplacian $L = D - W$. These basis functions are of size $|V| = n$, which can be problematic if n is large.

In the case of 3D objects, this can be as large as 10^5 or more. To address this, we will actually compute the Laplacian bases on subgraphs of much smaller size. In our experiments, we used the *normalized* Laplacian $\mathcal{L} = D^{-\frac{1}{2}}(D - W)D^{-\frac{1}{2}}$. A drawback of Laplacian approximation is that it detects only global smoothness, and may poorly approximate a function which is not globally smooth but only piecewise smooth, or with different smoothness in different regions (as in Figure 7.1). Diffusion wavelets were primarily designed to address these drawbacks.

7.3.2   Diffusion Wavelet Bases

Following the approach described in Chapter 4, we can construct multiscale diffusion wavelet bases by running the diffusion wavelet tree construction on a suitable diffusion operator. In the experiments in this chapter, we used the diffusion operator $T = D^{-\frac{1}{2}}WD^{-\frac{1}{2}}$.

A diffusion wavelet tree consist of orthogonal diffusion scaling functions Φ_j that are smooth bump functions, with some oscillations, at scale roughly 2^j (measured with respect to geodesic distance, for small j), and orthogonal wavelets Ψ_j that are smooth localized oscillatory functions at the same scale. The scaling functions Φ_j span a subspace V_j, with the property that $V_{j+1} \subseteq V_j$, and the span of Ψ_{j+1}, W_j, is the orthogonal complement of V_j into V_{j+1}. This is achieved by using the dyadic powers of the diffusion operator as "dilations", to create smoother and wider (always in a geodesic sense) "bump" functions (which represent densities for the symmetrized random walk after 2^j steps), and orthogonalizing and downsampling appropriately to transform sets of "bumps" into orthonormal scaling functions. The algorithm used is the one described in Chapter 4.

7.4   SCALING TO LARGE GRAPHS USING GRAPH PARTITIONING

We now describe one method of scaling the approach of adaptive mesh compression to large graphs—applicable to both Laplacian eigenvectors and diffusion wavelet bases. A natural divide-and-conquer strategy is to decompose the original graph into subgraphs, and then compute local basis functions on each subgraph. A number of graph-partitioning methods are available, including spectral methods that use the low-order eigenvectors of the Laplacian to decompose graphs, as well as hybrid methods that combine spectral analysis with other techniques. Karni and Gotsman [59] used the METIS system [60], which is a fast graph-partitioning algorithm that can decompose even very large graphs on the order of 10^6 vertices. METIS uses a *multiscale* approach to graph partitioning, where the original graph is "coarsened" by collapsing vertices (and their associated edges) to produce a series of smaller graphs, which are successively partitioned followed by uncoarsening steps mapping the partitions found back to the lower-level graphs.

7.4.1   Computing Local Basis Functions

Once a graph $G = (V, E, W)$ has been partitioned into a set of k partitions $G_i = (V_i, E_i, W_i)$, we compute a set of basis functions on each subgraph, either using the eigenvectors of the Laplacian, or multiscale local diffusion wavelet bases. One subtle issue is the boundary effects that can result from the fact that there are edges that lie in the intersection of two or more subgraphs. One approach to dealing with boundary effects is to use the *Dirichlet-* or *Neumann-* adjusted Laplacian matrices [26], which we do not address here.

7.5   MESH COMPRESSION USING FOURIER AND WAVELET BASES

In this section, we present a series of detailed experiments, evaluating the multiscale approach to the adaptive compression of 3D objects [76]. To compare the effectiveness of mesh geometry reconstruction by projection onto a set of Laplacian or diffusion wavelet bases, some notion of error needs to be defined. The most straightforward method is to compare the difference between the predicted mesh coordinates $\hat{v}$ with the true mesh coordinates v, that is the *geometric error* between two models M_1 and M_2 is defined as

$$\|M_1 - M_2\|_g = \sum_{v \in V} \sum_{i \in (x,y,z)} (\hat{v}_i - v_i)^2, \tag{7.1}$$

where, for example, $\hat{v}_x$ gives the approximated x coordinate and v_x is the exact known x coordinate of vertex v. Unfortunately, geometric error by itself is not sufficient, since it is possible that a model may be close geometrically, and yet provide a poor "visual" reconstruction. To deal with this, Karni and Gotsman [59] use a second metric, called the *geometric Laplacian*, defined as follows:

$$GL(v_i) = v_i - \frac{\sum_{j \in n(i)} l_{ij}^{-1} v_j}{\sum_{j \in n(i)} l_{ij}^{-1}}, \tag{7.2}$$

where $n(v)$ is the set of neighbors of vertex v, and v_i again is the ith index of the mesh coordinate geometry (for $i = x, y, z$). This term intuitively measures the difference between the prediction made by simply averaging the coordinates of the neighbors of a vertex versus the actual prediction. The final error in approximation is then defined as the sum of the normalized geometric Laplacian error and the geometric error:

$$\|M_1 - M_2\| = \frac{1}{2n} \left(\|M_1 - M_2\|_g + \sum_{v \in V} \sum_i GL(v_i) \right). \tag{7.3}$$

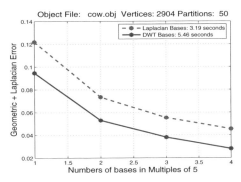

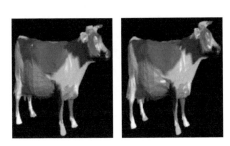

FIGURE 7.4: Comparison of Laplacian (top left) and diffusion wavelet (top right) approximation of a 3D "cow" model with $|V| = 2904$ vertices.

7.5.1   Compression of Small Objects

We first compare the mesh compression of global Laplacian eigenvectors against multiscale diffusion wavelet bases for "small" 3D objects, where by "small" we mean objects with mesh graphs of size $\leq 10^4$ vertices. Figures 7.4–7.7 compare the performance on a "cow", "camel", "pig", and "Max Planck" models, respectively. Each experiment was carried out using the same set of parameters. The overall graph was partitioned into 50 subgraphs, and then a varying number of basis functions were constructed on each subgraph. The errors introduced in each local subgraph mesh approximation were then added together to produce the final plots shown. In each graph, the horizontal axes measures the number of basis functions, and the vertical axes measures the sum of the geometric error and the geometric Laplacian error, as defined above. In each graph, the bottom curve represents multiscale diffusion bases, and the top curve

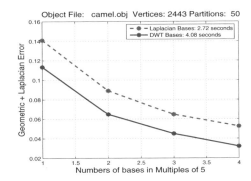

FIGURE 7.5: Comparison of Laplacian (top left) and diffusion wavelet (top right) approximation of a 3D "camel" model with $|V| = 2443$ vertices.

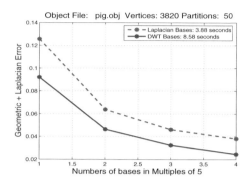

FIGURE 7.6: Comparison of Laplacian (top left) and diffusion wavelet (top right) approximation of a 3D "pig" model with $|V| = 3820$ vertices.

represents Laplacian bases. It is clear that the multiscale diffusion wavelet bases consistently perform better than the partitioned Laplacian eigenvector bases. The running times shown are the average time for a specific number of bases.[1]

7.5.2   Partition Size Versus Error

The divide-and-conquer approach seems a natural way to make the adaptive spectral compression problem more tractable, but it comes at a price. As the number of partitions grows, the error is likely to increase due to boundary effects, but the running time reduces.

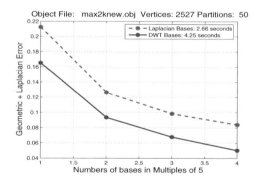

FIGURE 7.7: Comparison of Laplacian (top left) and diffusion wavelet (top right) approximation of a 3D model of Max Planck with $|V| = 2527$ vertices.

---

[1]Unlike the `eigs` package in MATLAB for computing eigenvectors, the diffusion wavelet code is not yet highly optimized.

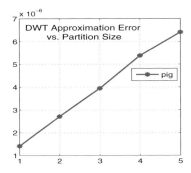
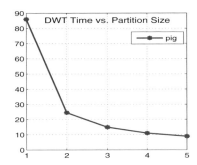

FIGURE 7.8: Comparison of average error and running times (seconds) over different partition sizes, showing error increases sub-linearly, but running time reduces super-linearly as number of partitions increase.

We explore this tradeoff for the "pig" model analyzed above. Figure 7.8 displays the change in error and running times versus partition size for the "pig" model for the diffusion wavelet model. Figure 7.9 shows the equivalent result for the Fourier bases. The increase in error turns out to depend on the smoothness of the model: the camel exhibits the worst increase in error whereas the pig exhibits the least. As shown in the figure, the error increase is at best linear, but the running time shows a significant super-linear decrease.

7.5.3   Compression of Large Objects

In this section, we compare the performance of multiscale diffusion bases against Laplacian bases on larger 3D objects, where the number of vertices $|V| > 10^4$. Specifically, Figure 7.10 compares multiscale diffusion wavelet bases versus global Laplacian bases on an "Elephant"

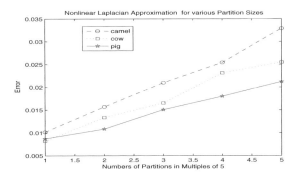

FIGURE 7.9: Comparison of error for different partition sizes for Laplacian eigenvector bases. As the partition size is increased, error increases as well due to boundary effects.

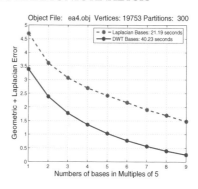
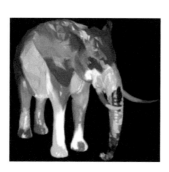

Object File: ea4.obj Vertices: 19753 Partitions: 300

FIGURE 7.10: Results for an "Elephant" model, a 3D object with 19 753 vertices and 59 053 edges.

model. The colors indicate the partitions on which local basis functions were computed. As in the earlier "cow" model, sharp features such as the tusks are rendered with much higher fidelity by the diffusion wavelet bases. Figure 7.11 plots the results for the "Stanford Bunny", a standard benchmark problem in computer graphics.

7.6   SUMMARY

This chapter explored the application of representation discovery to compression in 3D computer graphics. Increasingly, multimedia content on the web, in computer games, and in the next-generation of animated movies will rely on 3D graphics. Current Fourier and wavelet compression standards, such as JPEG and JPEG 2000, do not readily extend to 3D topologies. The challenges to be dealt with are the very large sizes of the graphs involved, the need for real-time compression, and the ability to compactly store and transmit the bases. Many extensions of this approach remain to be investigated. 3D models are sometimes specified by Euclidean point sets $\in \mathbb{R}^n$, which requires an additional graph construction phase such as that explored in

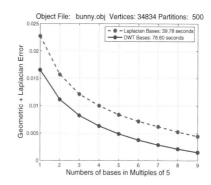

Object File: bunny.obj Vertices: 34834 Partitions: 500

FIGURE 7.11: Results for "Stanford Bunny", a 3D object with 34 834 vertices and 104 288 edges.

Chapter 6. The multiscale diffusion bases can also be modified to yield *geometry-aware* bases [107], where the coordinate function being approximated can influence the bases constructed.

7.7   BIBLIOGRAPHICAL REMARKS

The use of the graph Laplacian to approximate mesh geometry was pioneered by Karni and Gotsman [59], building on earlier work on Fourier descriptors by Taubin [112, 113]. The results in this chapter comparing diffusion wavelets and Laplacian eigenvectors is based on [76].

CHAPTER 8

Case Study: Natural Language

One of the most active areas in artificial intelligence is the statistical analysis of natural language, including information extraction and retrieval (IR) from large document corpora [83]. In this chapter, we describe the application of representation discovery to IR and learning from text. Specifically, we describe a new *wavelet-based* approach to the problem of learning hierarchical topic models from a given corpora of text documents. We begin with a classic well-studied method in this area: latent semantic indexing (LSI) [36] applies singular value decomposition (SVD) to extract structure from the *term–document matrix*, where the rows represent words, and the columns represent documents. This approach can be viewed as Fourier analysis applied to text, since in this case SVD finds eigenvector bases for the row space and column space of the term–document matrix.

The limitations of LSI derive from the intrinsic limitations of Fourier analysis. The analysis does not reveal multiscale regularities across documents. We show how diffusion wavelets can be applied to reveal multiscale regularities across documents. We also describe the use of diffusion wavelets to cluster documents by analyzing a diffusion process on the graph that reflects similarity across documents [123]. The key strength of the wavelet-based approach to topic discovery is that it can automatically determine the number of levels of the topic hierarchy of the corpora, as well as the number of topics at each level. Further, when the input term–term matrix is a "diffusion-like" operator, the diffusion wavelet algorithm runs in time approximately linear (within a logarithmic factor) in the number of nonzero elements of the matrix. We illustrate the approach on two real applications: a collection of NIPS papers and messages from an eBay Discussion forum.

8.1 INTRODUCTION

The problem of analyzing text corpora has emerged as one of the most active areas in data mining and machine learning for the World Wide Web. The goal here is to extract succinct descriptions of the members of a collection that enable efficient generalization and further processing. Topic models are an important tool because they are capable of identifying latent semantic components in unlabeled text data. Topic models have been successfully used to analyze text information on the web for many tasks. A topic can be viewed as a distribution on words. The "high-frequency" words that contribute more to each topic provide keywords that briefly summarize the main themes in a collection. Topic modeling discovers a set of topics expressed by documents, providing quantitative measures that can be used to identify the content of documents. Popularly used topic models include Latent Semantic Indexing (LSI) [36], probabilistic Latent Semantic Indexing (pLSI) [53], and Latent Dirichlet Allocation (LDA) [18].

Given a collection of text documents, it is important to extract structural information regarding the concepts/topics at *multiple* levels. A canonical example of topic discovery is the set of papers submitted to the International Conference on Neural Information Processing Systems or NIPS. Using the NIPS data set as an example, at the most abstract level, there are two main topics in the published papers: machine learning and neuroscience. Researchers who submit to the NIPS conference usually need more detailed information, such as which topic is more popular, or whether there are some new research topics that they should pay attention to. At the next level, there may be topics pertaining to a number of areas. In summary, the problem of hierarchical topic modeling can be formalized as follows: given a collection of documents, each of which contains a bag of words, can we discover common topics in the documents and organize these topics into a hierarchy?

In this chapter, we explore the application of Fourier and wavelet bases to automatically extracting hierarchical topics from a given corpus. In particular, we show that diffusion wavelets provide a novel way of extracting multiscale structure across text documents [123]. The key strength of this approach to topic discovery is that it can automatically determine the number of levels of the topic hierarchy of the corpora, as well as the number of topics at each level. Further, when the input term–term matrix is a "diffusion-like" operator, the diffusion wavelet algorithm runs in time approximately linear (within a logarithmic factor) in the number of nonzero elements of the matrix (modulo a large constant factor). We illustrate the diffusion wavelet approach on two real applications: a collection of NIPS papers and messages from an eBay Discussion forum.

The wavelet approach automatically reveals the geometric structures of the document collection at different scales, and offers the following advantages. (1) It is a model-free, data-driven, and mostly parameter-free method. It automatically generates both topic hierarchies as well as the topics at each level. The only input information is the term–document matrix and a resolution parameter. (2) In contrast to the topic vectors learned from latent semantic indexing, the scaling functions in diffusion wavelets have local support. This sparsity is particularly useful when the concept only involves a small group of words. (3) The topic hierarchy is not a tree structure. Topic vectors at adjacent levels do not have parent–child relationships. It is well known that tree-structured models have some limitations. For example, documents that are in distinct subsets of a corpus might share a topic, which is hard to model using a tree structure.

We test the wavelet approach on two real-world data sets: the NIPS paper data set [100] and a data set collected from an eBay^{TM} Computers, Networking and IT discussion board [43]. The results show that the diffusion wavelet-based method can automatically identify the structure of the collection at different scales, and the topics learned from each level nicely capture semantically meaningful categories.

8.2   FOURIER ANALYSIS OF TEXT

We begin by introducing the term–document matrix representation, and then review latent semantic indexing (LSI) [36], which is a classic Fourier-type analysis of text based on eigenvectors.

8.2.1   Term–Document Matrix Representation

In a collection of documents (defined on a vocabulary with n terms), any document can be represented as a vector in $\mathbb{R}^n$, where each dimension represents a term. The ith element of the vector can be some function of the number of times that the ith term occurs in the document. There are several possible ways to define the function to be used here (frequency, term frequency inverse document frequency (TFIDF), etc), but the precise method is not important. Let A be an $n \times m$ matrix of rank r whose rows represent terms and columns represent documents, with singular value decomposition $A = U\Sigma V^T$. Let the singular values of A be ordered as $\delta_1 \geq \delta_2 \geq \cdots \geq \delta_r$.

Each row of A is a vector corresponding to a term, giving its relation to each document. Likewise, each column of A is a vector corresponding to a document, giving its relation to each term. The matrix AA^T defines an inner product between any two term vectors, and gives the correlation between terms over the documents. From Chapter 2 and basic linear algebra, we know $AA^T = (U\Sigma V^T)(U\Sigma V^T)^T = U\Sigma\Sigma^T U^T$, so the column vectors of U are the eigenvectors of AA^T.

Let us define a new matrix $W = AA^T$. Obviously, the term–term matrix W is a *Gram* matrix (see Chapter 2 for a review of Gram matrices) with non-negative entries. Let D be a diagonal matrix, where D_{ii} is the sum of the entries on the ith row of W. Then, the *normalized Laplacian operator* (discussed in detail in Chapter 3) associated with W is $\mathcal{L} = I - D^{-0.5}WD^{-0.5}$ [27]. We define T as $D^{-0.5}WD^{-0.5}$, which is the normalized term–term matrix.

8.2.2   Latent Semantic Indexing

Latent semantic indexing (LSI) [36] applies singular value decomposition (SVD) to topics in a text corpus. A brief overview of SVD was provided in Chapter 2. The key idea is to map high-dimensional vectors to a lower-dimensional eigenvector representation, which captures latent semantic space. The goal of LSI is to find a mapping that reveals semantic relations between the entities of the interest. LSA is a "flat" topic model, which means it cannot find hierarchical topics.

The singular value decomposition of the term–document matrix A is $A = U\Sigma V^T$, where $\Sigma = \text{diag}(\delta_1, \ldots, \delta_r)$, U is an $n \times r$ matrix whose columns are orthonormal, and V is an $m \times r$ matrix whose columns are also orthonormal. LSI constructs a rank-k approximation of the matrix by keeping the k largest singular values in the above decomposition, where k is usually much smaller than r. More precisely, the best rank-k approximation is given by $A_k = U_k\Sigma_k V_k^T$, and it can be shown that this approximation has the smallest error (w.r.t. the Frobenius norm) [47]. In LSI, the columns of $\Sigma_k V_k^T$ are used to represent the documents in a space spanned by the columns of U_k. The space can be called LSI space of A. Each of the column vectors of U_k is related to a concept, and represents a topic in the given collection of documents.

8.3   MULTISCALE ANALYSIS OF TEXT USING DIFFUSION WAVELETS

We now illustrate how diffusion wavelets can be applied to text analysis. The diffusion-based approach uses the diffusion *scaling functions* constructed by diffusion wavelets [30]. The construction of diffusion wavelets was described in Chapter 4. Diffusion wavelets can be interpreted geometrically as projecting data to a lower-dimensional space by using the scaling functions while preserving the large-scale information inherent in the data. The projections provide multiscale embedding, which means they automatically reveal the geometric structure of the data at different scales. The subspace spanned by scaling functions learned from T is in fact the subspace spanned by certain eigenvectors of $\mathcal{L}$ (with smallest eigenvalues) up to a precision ε [30]. For the normalized Laplacian, the eigenvectors of T and $\mathcal{L}$ are exactly the same, so instead of learning eigenvectors of T, we can learn diffusion scaling functions from T since they are the basis functions that effectively also span the space spanned by the singular vectors used in LSI.

The diffusion-model approach is completely data-driven: the user only needs to provide the desired precision, all the remaining computation is done automatically. This includes the identification of the number of the concept levels (topic levels), basis functions (topic vectors) at each level, and new representations of the set of documents at each new level. It can be shown that when the relationships between examples (for this case, relationship between terms) are characterized by "diffusion-like matrices" (matrices whose high power is of low numerical rank), computation of the basis functions can be done in approximately linear time [72]. This is in contrast to the computation of k eigenvectors, which is approximately $O(kn^2)$. The diffusion model combines the concept of multiscale representation and a modified QR decomposition.

8.3.1   Fourier Versus Wavelet Analysis of Text

As discussed in Section 8.2.2, latent semantic indexing (LSI) is based on modeling topic vectors from the given document collection as the eigenvectors from the normalized term–term matrix T. However, there are two problems when we apply this technique in practice. One problem is that LSI only learns "flat" topics, i.e. all the topics are at the same level, since the eigenvectors of T and T^i are the same. Another problem is that the computation of k eigenvectors of a $n \times n$ matrix is in general a $O(kn^2)$ time task, which can be expensive for large n.

The diffusion wavelet approach, in contrast, uses multiscale basis functions learned via the computation of diffusion wavelets. From a numerical point of view, they correspond to QR decompositions of powers of $T = I - \mathcal{L}$. The QR decomposition of a matrix T, which was reviewed in Chapter 2, decomposes the matrix into an orthogonal matrix (Q) and a triangular matrix (R), where $T = QR$. Columns in Q are orthogonal basis functions spanning the same space as columns in the matrix T. Here, R can be thought as the new representation of T with respect to the space spanned by the columns (basis functions) of Q. A well-known approach for QR decomposition is the Gram–Schmidt orthogonalization, but there are a variety of other methods [47].

8.3.2   Main Algorithm

Figure 8.1 describes the precise steps needed to learn a hierarchical topic model using multiscale diffusion wavelet analysis. This approach assumes the term–document matrix A (defined in Section 8.2.2) is already given.

8.3.3   Advantages of the Diffusion Wavelet Approach

Fewer Parameters.   As illustrated in Chapter 4, the spaces at different levels of a diffusion wavelet tree are spanned by different numbers of basis functions. These numbers reveal the dimensions of the relevant geometric structures of data at different levels. These numbers are completely data-driven, so instead of requiring a user to input the number of levels, number of topics, etc, the wavelet approach can automatically determine the structure of the hierarchy and simultaneously generate the topics at each level. In fact, once the term–document matrix A is given, a user only needs to specify one parameter ε—the precision. If the precision is high, the algorithm needs more time to run, since the diffusion process is slower, and vice versa.

Computational Complexity.   The diffusion wavelet tree algorithm runs in time linear within a logarithmic factor when T is a "diffusion-like" matrix, as shown in [30]. The main idea is that most examples defined in the "diffusion-like matrix" have "small" degrees in which transitions

1. **Construct the normalized term–term matrix**: $T = D^{-0.5} AA^T D^{-0.5}$, where A is the term–document matrix, AA^T is the term–term matrix, D is a diagonal matrix where D_{ii} is the sum of the entries on the ith row of AA^T.

2. **Optional Sparsification Step**: Sparsify T to make it more "diffusion-like" by keeping the largest k entries in each row of AA^T, and setting all the other entries to zero (see Section 8.3.3).

3. **Generate Diffusion Model**: Run the diffusion wavelet procedure described in Figure 4.4 in Chapter 4.

$$\{\phi_j, \psi_j\} = \text{DWT}(T_0, \phi_0, QR, J, \varepsilon)$$

   - The initial diffusion operator, $T_0 = T$, is the normalized term–term matrix represented on the unit vector basis ϕ_0.
   - QR is a modified QR decomposition used in the diffusion wavelet construction, as described in Chapter 4 (see also [30]).
   - J is the number of desired levels. If J is omitted, the diffusion wavelet tree algorithm will run until the representation of T^{2^j} converges to a scalar (the leading eigenvalue of T) at the topmost level, at which point the construction will terminate.
   - ε is the desired numerical precision used in the diffusion wavelet algorithm.
   - ϕ_j is the resulting set of diffusion scaling functions at level j.
   - ψ_j is the resulting set of wavelet functions at level j.

4. **Compute extended basis functions**:

   - The representation of the basis functions from level j on the original space $[\phi_j]_{\phi_0}$ is computed as follows:

$$[\phi_j]_{\phi_0} = [\phi_j]_{\phi_{j-1}} [\phi_{j-1}]_{\phi_{j-2}} \cdots [\phi_1]_{\phi_0} [\phi_0]_{\phi_0}$$

   - $[\phi_j]_{\phi_0}$ is a $n \times n_j$ matrix. Each column vector represents a topic at level j whose kth entry is the kth term's contribution to this topic at level j.

5. **Apply the extended basis functions:**

   - Construct topic hierarchy using each column vector of $[\phi_j]_{\phi_0}$, which specifies a topic at level j.
   - Project documents represented in the original space on the subspace spanned by extended basis functions from different levels and construct multilevel representations of each document.

FIGURE 8.1: Algorithm for constructing a multiscale diffusion wavelet representation of a collection of text documents.

are allowed only among neighboring points, and the spectrum of the transition matrix decays rapidly [30]. This result is in contrast to the time needed to compute k eigenvectors, which is $O(kn^2)$.

In many applications, the normalized term–term matrix T is already a "diffusion-like matrix". If it is not, we can use the following procedure to convert it to such a matrix. The basic idea is that for each term in the collection, we only consider the most relevant k terms, since the relationships between terms that co-occur many times are more important. The same technique has been used in manifold learning [101] to generate the relationship graph from the given data examples. This sparsification algorithm keeps the top k entries in each row of AA^T, and sets all the other entries to zero. The resulting matrix is not symmetric, so we need to symmetrize it at the end.

The Topic Vectors have Local Support.   Given the normalized term–term matrix T, the space spanned by the topic vectors are the same as the space spanned by some topic vectors learned from Latent Semantic Indexing up to a precision ε. However, the topic vectors (in fact eigenvectors) from LSI have a potential drawback in that they only detect global smoothness, and may poorly model the concept/topic, if they are not globally smooth but only piecewise smooth, or have different smoothness in different regions. Unlike the global nature of eigenvectors, the topic vectors from diffusion models are local, and better capture some concepts/topics that only involve a particular group of words.

Hierarchical Topic Structure.   A natural representation for a hierarchical topic model is to organize the topics into a tree. For example, a well-known hierarchical topic model is hLDA [17], where each document is assigned to a path through the topic tree, and each word in a given document is assigned to a topic at one of the levels of that path. The tree structure has some limitations. One problem is that it is very important to identify the correct tree. In order to learn such a tree, for example, hLDA applies the so-called nested Chinese restaurant process [17]. Another problem is that documents sharing the same topic might be in quite different subsets of a corpus. This is hard to model with tree structures. The multilevel structure from the diffusion wavelet tree is not based on a tree structure. In other words, there is no such path that goes from the root to a leaf node. We generate topic vectors (basis functions) at different levels, but for each topic vector, there is no parent vector at the upper level. Topics at different levels are "independent".

Scalability.   The complexity of generating a diffusion model mostly depends on the size of the vocabulary in the corpus, but not the number of the documents, or the number of the tokens.

We know no matter how large the corpus is, the size of the vocabulary set is determined, and we can always set a threshold to filter terms that only appear a small number of times. So the diffusion wavelet approach can be scaled to large data sets.

8.4   EXPERIMENTAL RESULTS

In this section, we describe the results of the diffusion wavelet-based approach to topic discovery using two real-world data sets. Since this approach is largely parameter-free, we do not need any special settings. The precision parameter used in all the experiments was set at $\varepsilon = 10^{-5}$. One problem that is important but we have not addressed so far is how to interpret topics learned from diffusion models. Any given topic vector v is a column vector of length n, where n is the size of the vocabulary set. The entry $v[i]$ represents the contribution of the term i to this topic. To illustrate the main concepts of the topic v, we sort the entries on v and print out the terms corresponding to the top ten entries. These terms should summarize the topics in the collection.

8.4.1   NIPS Paper Data Set

We generated hierarchical topics from the NIPS paper data set [100], which includes 1740 papers. The original vocabulary set has 13,649 terms. The corpus has 2,301,375 tokens in total. We filtered out the terms that appear ≤ 100 times in the corpus, and only 3413 terms were kept. The number of remaining tokens was 2,003,017. We performed two tests using the data.

8.4.2   Diffusion Model: Test 1

In Test 1, we follow the procedure described in Section 8.3. Running the diffusion wavelet algorithm results in a tree with five levels; the number of topics at each level is shown in Table 8.1. At the first level, each column in T is treated as a topic. At the second level, the

TABLE 8.1: Number of Topics at Different Levels (Diffusion Model, NIPS Test 1)

LEVEL	NUMBER OF TOPICS
1	3413
2	1739
3	1052
4	37
5	2

TABLE 8.2: All Two Topics from Level 5 (Diffusion Model, NIPS Test 1)

TOPIC ID	TOP TEN TERMS
Topic (5,1)	Network learning model neural input data time function figure set
Topic (5,2)	Cells cell neurons firing cortex synaptic visual stimulus cortical neuron

TABLE 8.3: All 37 Topics from Level 4 (Diffusion Model, NIPS Test 1)

TOPIC ID	TOP TEN TERMS
Topic (4,1)	Network learning model neural input data time function figure set
Topic (4,2)	Cells cell neurons firing cortex synaptic visual cortical stimulus response
Topic (4,3)	Policy state action reinforcement actions learning reward MDP agent Sutton
Topic (4,4)	Mouse chain proteins region heavy receptor protein alpha human domains
Topic (4,5)	Distribution data Gaussian density Bayesian kernel posterior likelihood EM regression
Topic (4,6)	Chip circuit analog voltage VLSI transistor charge circuits gate cmos
Topic (4,7)	Image motion images object eye visual velocity chip vision face
Topic (4,8)	Speech hmm word speaker phonetic recognition spike Markov mixture acoustic
Topic (4,9)	iiii border iii texture ill bars suppression ground bar contextual
Topic (4,10)	Face facial images faces image tangent spike object views similarity
Topic (4,11)	Adaboost margin boosting classifiers head classifier hypothesis training SVM motion

(cont.)

TABLE 8.3: (*Continued*)

TOPIC ID	TOP TEN TERMS
Topic (4,12)	Dominance ocular orientation cortical development bands lgn lateral striate cortex
Topic (4,13)	Stress syllable song heavy linguistic vowel languages primary harmony language
Topic (4,14)	Motor control muscle arm controller inverse movement iiii trajectory kawato
Topic (4,15)	Hint hints monotonicity mostafa abu market schedules trading financial monotonic
Topic (4,16)	Sound auditory localization spectral sounds cochlear cue cues EEG frequency
Topic (4,17)	Obs obd pruning Hessian stork retraining pruned weight weights stress
Topic (4,18)	Routing traffic load shortest paths route path node message recovery
Topic (4,19)	Spike spikes motion trains noise rate stress spiking time timing
Topic (4,20)	Tangent distance prototypes simard transformations Euclidean rotation character vectors prototype
Topic (4,21)	EEG ICA artifacts locked blind sources separation component components independent
Topic (4,22)	Clause phrase parsing sentences obs parse query documents sentence harmony
Topic (4,23)	Obs theorem threshold gates maass polynomial bounds functions rational face
Topic (4,24)	Instructions instruction scheduling schedule dec blocks execution schedules block processor
Topic (4,25)	Student teacher overlaps queries saad face biases generalization facial documents
Topic (4,26)	vor head vestibular eye reflex cerebellum ocular spike velocity gain

(*cont.*)

TABLE 8.3: (*Continued*)

TOPIC ID	TOP TEN TERMS
Topic (4,27)	Oscillators oscillator oscillatory obs oscillation oscillations synchronization phase coupling wang
Topic (4,28)	Harmony tree smolensky parse trees student legal grammar child tensor
Topic (4,29)	Actor critic pendulum tsitsiklis pole barto harmony signature routing instructions
Topic (4,30)	Documents query document retrieval queries words relevant collection text ranking
Topic (4,31)	Classifier classifiers clause knn rbf tree nearest neighbor centers classification
Topic (4,32)	Stack symbol strings grammars string grammar automata grammatical automaton giles
Topic (4,33)	Song template production kohonen syllable pathway harmonic nucleus lesions motor
Topic (4,34)	Rat head place direction spike navigation dominance food card sharp
Topic (4,35)	Som gtm latent date map organizing parity kohonen manifold quantization
Topic (4,36)	hme experts expert tangent gating growing tree mixtures Jacobs distance
Topic (4,37)	Object views objects eeg adaboost view edelman instantiation viewpoint rigid

number of the columns is almost the same as the rank of T. At level 4, number of topics goes down to a reasonable number 37. Finally, at level 5, the number of topics is 2.

The two topics at level 5 are shown in Table 8.2. Topic 1 is related to machine learning, while topic 2 is related to neuroscience. These two topics are a reasonable way to partition the space of all papers that appear in the NIPS conference. The 37 topics at level 4 are shown in Table 8.3. Almost all these topics seem "meaningful".

TABLE 8.4: Number of Topics at Different Levels (Diffusion Model, NIPS Test 2)

LEVEL	NUMBER OF TOPICS
1	3413
2	2622
3	1180
4	136
5	22
6	2

8.4.3   Diffusion Model: Test 2

In Test 1, the diagonal elements of the normalized term–term matrix T are always large and represent the co-occurrence of a term and itself. This might not make perfect sense. So in Test 2, we set the diagonal elements to 0 and renormalized T. The new matrix might not be a "diffusion-like matrix", so we applied the method described in Section 8.3.3 to first convert it to a "diffusion-like matrix". Then the regular diffusion model is used to retrieve hierarchical topics.

The diffusion wavelet model identifies six levels of topics, and the number of topics at each level is shown in Table 8.4. At the first level, each column in T is treated as a topic. Then the number of topics at each level decreases till finally at level 6, there are two topics.

The two topics at level 6 are shown in Table 8.5. As with the results of Test 1, the two topics are related to machine learning and neuroscience. The 22 topics at level 5 are shown in

TABLE 8.5: All Two Topics from Level 6 (Diffusion Model, NIPS Test 2)

TOPIC ID	TOP TEN TERMS
Topic (6,1)	Network learning model neural input data time set function figure
Topic (6,2)	Voltage cells cell firing synaptic circuit cortex synapses cortical membrane

TABLE 8.6: All 22 Topics from Level 5 (Diffusion Model, NIPS Test 2)

TOPIC ID	TOP TEN TERMS
Topic (6,1)	Network learning model neural input data time set function figure
Topic (6,2)	Voltage circuit chip transistor cells cell synapse synaptic synapses transistors
Topic (6,3)	Ocular dominance eye orientation cortex cortical cells mouse head visual
Topic (6,4)	Mouse chain proteins heavy alpha region protein receptor domains human
Topic (6,5)	Policy action actions reinforcement reward sutton agent policies MDP Singh
Topic (6,6)	Word speech phonetic speaker phoneme speakers sentences hmm spoken letter
Topic (6,7)	Instructions instruction scheduling dec schedule blocks execution schedules hints hint
Topic (6,8)	Stack grammar grammars strings symbol string symbols giles automata grammatical
Topic (6,9)	Hint hints monotonicity mostafa abu market financial monotonic trading stock
Topic (6,10)	Tangent rotation distance transformations simard digit prototypes rotations rotated digits
Topic (6,11)	Motor cerebellum movement arm movements cerebellar muscle vor muscles command
Topic (6,12)	Adaboost margin boosting smola svm sch sv support vapnik svms
Topic (6,13)	ICA Blind separation sources EEG artifacts mixing source kurtosis independent
Topic (6,14)	Student teacher overlaps saad documents biases queries query symmetric document

(cont.)

TABLE 8.6: (*Continued*)

TOPIC ID	TOP TEN TERMS
Topic (6,15)	Tensor role roles binding product representation structures connectionist representations distributed
Topic (6,16)	Tangent calcium soma dendrite conductance dendritic distance simard somatic digit
Topic (6,17)	Posterior Bayesian mixture likelihood em experts ylx Gaussian hme kullback
Topic (6,18)	Road vehicle autonomous driving lane navigation land obstacle color food
Topic (6,19)	Obs obd pruning retraining stork Hessian pruned damage remove brain
Topic (6,20)	Oscillators oscillator oscillatory oscillation synchronization wang attractors oscillations spiral coupling
Topic (6,21)	Routing traffic shortest load paths route service call calls link
Topic (6,22)	Learning network neural time set input figure function training model

Table 8.6. Again, almost all these topics are reasonable. They nicely capture the function words in the corpus. There are 136 topics at level 4; we checked these topics, and found over 90% of them to be meaningful. We list 20 of them in Table 8.7. This test also confirms that the above described method of converting a regular term–term matrix to a diffusion-like matrix works reasonably well.

8.4.4   Running Times for Various Approaches

Given the collection with 2,003,017 tokens, in both of our diffusion model settings, we need roughly 15 min (2G PC with 2G memory) to do the multiscale analysis. This includes data preparation, construction of the diffusion wavelet tree and computing topic vectors at all the levels (five levels for test 1 and six levels for test 2). In contrast, it took 6 min to compute 37 topics using LDA on the same machine. However, LDA only computes a single-level "flat" topic model.

TABLE 8.7: Twenty Selected Topics from the 136 Topics at Level 4 (Diffusion Model, NIPS Test 2)

	TOP TEN TERMS
1	Kullback leibler ylx logarithmic divergence weighting pool factors yang experts
2	Arm inverse trajectory muscle kawato torque force kinematics movement workspace
3	sv sch smola kernels support kernel svm vapnik machines regularization
4	iiii border iii ill bars ground texture suppression bar figures
5	Monte Carlo posterior Bayesian hyperparameters neal mackay prior sampling Gaussian
6	Insect feeding food animal behaviors chemical behavior motivated begins Weiss
7	Lagrange multipliers constrained constraint constraints multiplier optimization differential permutation gold
8	Color land blue grey red green illumination surround light scenes
9	Fitness genetic evaluations Holland climbing population hill outperform codes royal
10	Vertices graphs graph vertex clique maximal matching edges Hopfield planar
11	Rotations lie rotation law rotated angles cos plane group sphere
12	Linsker eigenvector eigenvectors eigenvalue centre eigenvalues Miller principal regime largest
13	Similarity probe analogy dot spot causing products structural alignment relations
14	Magnetic sensor hall fusion sensing receptors devices sensors flow device
15	Leaf tree leaves branching trees growth suffix splits split branch
16	Wavelet coefficients transform histograms transforms video gamma marginal filters basis
17	Planning plan robot goal world dyna thrun exploration environment experience
18	Stability Lyapunov equilibrium bifurcation stable attention stream chaos Jones attentional
19	EM density estimation likelihood maximization expectation dempster spline densities mixture
20	Manifold interpolation lip-dimensional gtm appearance embedded dimensionality surface grid

TABLE 8.8: Number of Topics at Different Levels (Diffusion Model, eBay I.T. Board)

LEVEL	NUMBER OF TOPICS
1	2005
2	1903
3	202
4	10
5	4
5	2

8.4.5   eBay Discussion Forum Data Set

The second real data set we tested is "Computers, Networking & I.T. Discussion Board" at eBay [43]. This forum is created specifically for members of the eBay Computers, Networking & I.T. community. People can share their views and suggestions in this forum.

We selected all the posts from this board on Aug 10th, 2007. This generated a data set with 8641 posts. There are more than 20,000 terms in the vocabulary set. We only kept the terms that occurred >20 times in the collection, which resulted in a vocabulary set with 2005 terms. There are 211, 691 tokens in this set. We follow the procedure described in Section 8.3. Our diffusion model identifies six levels of topics, and the number of topics at each level is shown in Table 8.8.

In such discussion boards, there are usually many topics being discussed at any time. The 202 topics at level 3 seems a reasonable number to model the concepts that best follow human intuition. We manually checked the topics, and confirmed that more than 90% of these topics are meaningful. They nicely captured the function words in the corpus. The topics from levels 4–6 are more general, but they are described in such a high-level way and are not very interesting. To save the space, we show in Table 8.9 only 30 topics from level 3.

8.5   CONCLUSIONS

In this chapter, we explored the application of representation discovery to the statistical analysis of text. In particular, we applied diffusion wavelets [30, 123] to learn hierarchical topics from a given corpora of text documents. Unlike Fourier-based methods, such as latent semantic indexing, the wavelet approach can automatically determine the number of topic levels in the

TABLE 8.9: Thirty Selected Topics from the 202 Topics at Level 3 (eBay I.T. Board)

	TOP TEN TERMS
1	press enter disk xp boot restart drive key installation unit
2	songs ipod mp3 itunes player library napster song wmp play
3	cartridge cartridges toner printer ink laser printers print printing hp
4	iphone phones apple stores demand iphones press friday calls store
5	refresh rate lcd crt monitor image display monitors rates effect
6	fb neg truth leave dont smart positive thebay buyer useless
7	hd refresh partition defrag category rate tests xl os floppy
8	pc100 pc133 ecc memory mb sticks mhz asus psu cpu
9	avg norton virus spyware av spybot aware majorgeeks anti popup
10	garmin maps gps road street car liked wife confused directions
11	clock cache mhz intel bus properties kb cpu acpi size
12	paypal protection payment credit union pp western fee shipping account
13	crt lcd monitor monitors ecc resolution space concern hooked tower
14	router wireless dsl linksys connection modem connect network defrag cable
15	cmos bios ecc zone battery lil jumper clock floppy doa
16	usps ups shipping label package weight box ship size print
17	mode safe bar nasty onetouch data icons move status loaded
18	safari opera update annoying itunes apple updates macs corner junkin 423
19	refurbished condition seller unit laptop warranty government xps sealed units
20	cingular provider mobile phones cell verizon country network addition features
21	mac apps g5 apple written marketing refurbished virtual bose testing
22	cartridges laser printers cost toner research bump scanner outlet cheaper

(cont.)

TABLE 8.9: (*Continued*)

	TOP TEN TERMS
23	mb memory physical swap total transfer ann ram file dsl
24	ipod itunes library external mini movies sync dvd menu license
25	blocker popup av spam google popups everytime oe personally safari
26	union western payment mhz send taxes font refurbished delivered adobe
27	pdf cant edit contents scanning document values scan clicked uninstall
28	image resolution screen perk paint manually ghost select lazy database
29	watt atx p4 psu connected unit em transfer agp file
30	wife difficult solve mine logic fun daughter refund cookies town

corpora and the topic vectors at each level. When the input term–term matrix is "diffusion-like", the wavelet construction algorithm can be done in approximately linear time (albeit with a large constant factor overhead). Experiments on a NIPS conference paper data set and an eBay discussion board data set show that the multiscale wavelet approach successfully extracts hierarchical regularities at multiple levels, which form semantically meaningful topics. The same approach can be used in many other applications in information retrieval, such as finding document representations at different levels, and clustering of documents.

8.6 BIBLIOGRAPHICAL REMARKS

An overview of latent semantic indexing is given in [36]. Parametric graphical models have been studied extensively in text learning, such as Latent Dirichlet Allocation [18]. This material in this chapter is based on [123]. I am indebted to Chang Wang for running the experiments described in this chapter.

CHAPTER 9

Future Directions

Representation discovery is an actively developing area of research, with many directions yet to be explored. We have described a mathematically principled approach to representation discovery, based on applying the ideas of harmonic analysis. Yet, this is but one approach, and many other approaches remain to be explored. Staying within this framework, in this chapter we summarize a few promising directions for future research. First, we summarize an active area of research called *compressed sensing* [22]. A intriguing result shows that it is possible to (exactly) recover a sparse signal from a set of *incoherent dual bases*. For example, instead of computing the complete Fourier transform of a signal, it is sufficient to compute a small number of random coefficients, and reconstruct the original signal using $\mathbb{L}_1$-norm minimization. There are a wealth of representations used in AI not discussed in this book, in particular rich representations such as logic. Another direction for future work is to construct factored Fourier bases that exploit the representational power of propositional and first-order logic [54, 13]. Finally, we briefly summarize group representation theory [51, 105], an overarching framework that unifies both Fourier and wavelet representations, and provides some broad principles for constructing compact representations in homogeneous spaces.

9.1 COMPRESSED SENSING

Compressed sensing is one of the most active recent areas in computational harmonic analysis [22]. To help understand this approach, let us revisit the basic framework of harmonic analysis in terms of its use in compression. Compression of images using JPEG (and JPEG 2000) [122] is based on the familiar *analysis–synthesis* perspective of reconstructing a vector v from a set of measurements $c_i = \langle v, \phi_i \rangle$, which are then linearly combined with the dual basis $\sum_i c_i \psi_i$. JPEG uses the discrete-cosine transform (DCT) [3], which is a Fourier basis for 2D images, whereas JPEG 2000 uses a wavelet basis. Both of these approaches work because most natural images are highly compressible: for example, often 95% or more of the coefficients in a wavelet representation of a natural image can be discarded without any noticeable visual loss! This implies that natural images are highly sparse, not in the original pixel basis, but in a Fourier or wavelet basis. This phenomena is not restricted to visual images, but in fact holds for a large class of applications including speech, medical imaging devices such as CAT and MRI scanners, and bioinformatics (gene assay chips).

The core idea underlying compressed sensing is remarkably simple: if the vector v is sparse in some basis B, why is it even necessary to compute all the coefficients in the basis in order to extract a sparse representation? For example, why do digital cameras record images using millions of pixels (thus constructing a representation that uses millions of measurements and resulting coefficients), only to throw away almost all this information when converting the pixel representation to a JPEG (or JPEG 2000) basis? Is it possible to make a much smaller (say a few thousand) set of measurements on the input image (or signal) and be able to reconstruct it exactly? Even more interestingly, can these measurements be made *non-adaptively*?

Compressed sensing show that signals can be recovered (sometimes exactly) from a small set of random non-adaptive measurements, as long as the original signal is sparse with respect to some basis [22]. Figure 9.1 illustrates the main idea. Here, the original vector v is of length 1024, with only around 20 nonzero entries. This vector was deliberately chosen to be sparse in the unit vector bases for simplicity. Shown below is an almost exact reconstruction of the original vector, from only 128 random measurements. The measurements in this case were randomly sampled from rows of the discrete-cosine transform (DCT) matrix. It is important to stress that the sampling of the rows of the DCT matrix was not based on knowledge of

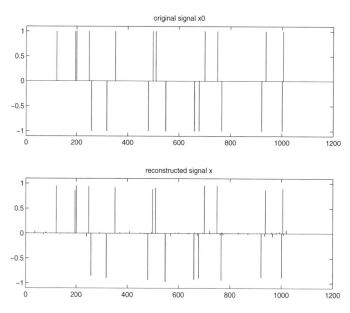

FIGURE 9.1: An example of compressed sensing, showing that sparse signals (top) can be recovered almost exactly (bottom) from a small number of *non-adaptive* random measurements.

where the signal is nonzero. The measurements were completely non-adaptive. Compression essentially implies using a smaller set of bases, instead of the full set, where $M \ll N$ linear measurements are made of the form:

$$v \Rightarrow \{\langle v, \phi_1 \rangle, \ldots, \langle v, \phi_M \rangle\}.$$

Define the *sensing matrix* Φ to be an $M \times N$ matrix, where we typically assume that $M \ll N$. For example, in Figure 9.1, $M = 128$, but $N = 1024$. The reconstructed signal u^* is the minimizer of the $\mathbb{L}_1$-norm, subject to the data constraints specified as

$$v^* = \min_v \|v\|_1, \quad \text{subject to} \quad \Phi v = u.$$

A central result proved by Candes and Tao [22] is that Φ can be chosen in a way that does not depend on the exact knowledge of the sparsity structure of v (either in its original basis or some other bases like a Fourier or wavelet basis). For example, if the rows of Φ are randomly chosen Gaussian-distributed vectors (thus, the matrix consists of N samples of an M-dimensional sphere), then there is a constant C such that if the support of v has size K and $M \geq CK \log(\frac{N}{K})$, then the solution of the above $\mathbb{L}_1$-norm minimization will be the original vector $v^* = v$ with high probability.

Compressed sensing is an intriguing new development in sparse representation theory, which promises to lead to new algorithms for both sensing as well as reconstruction. Applications of this theory are currently being explored in a large number of domains, and research on the implications of compressed sensing for representation discovery need to be fully investigated.

9.2   HARMONIC ANALYSIS AND LOGIC

In this book, we have largely focused on representations that are derived from vector spaces. A natural question that may arise in the mind of an AI reader is whether the harmonic analysis approach can be extended to the sort of "rich" representations used in AI, including propositional and first-order logic. Happily, the answer to this question is yes, and we briefly summarize the use of Fourier representations in propositional logic. As it turns out, Fourier representations have largely displaced the more common local *state-table* representation in many areas, such as VLSI design [118] where fast spectral methods have been devised to transform boolean functions into the Fourier domain. Even more strikingly, researchers in computational learning theory have discovered that Fourier bases make it possible to provably learn boolean functions more reliably from data than other more commonly used representations [54].

Figure 9.2 contrasts the difference between the global Fourier representation of a simple boolean function with the more conventional truth-table representation. Note the stark contrast in the two representations: the truth-table representation specifies the value of the function *locally*, whereas each coefficient in the Fourier representation specifies the function *globally*.

$f = x_1 \lor x_3 \lor \neg x_2 x_3$

x_1	x_2	x_3	f	$\hat{f}$
0	0	0	0	4
0	0	1	1	0
0	1	0	0	2
0	1	1	0	−2
1	0	0	1	−2
1	0	1	1	−2
1	1	0	1	0
1	1	1	0	0

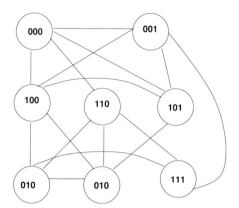

FIGURE 9.2: Boolean functions in propositional logic can be represented in a global Fourier basis. The Fourier coefficients $\hat{f}$ are the eigenvalues of the graph shown.

How are the Fourier coefficients derived from the truth table representation? There are several approaches, but a simple way to understand the Fourier coefficients is to view them as the eigenvalues of a graph derived from the truth table representation of the boolean function [13]. Here, each row m_i of the truth table defines a vertex in the graph. There is an edge from the vertex m_i to the vertex m_j if $f(m_i \oplus m_j) = 1$, where $\oplus$ represents the exclusive-or of the bit patterns in the rows m_i and m_j.

Obviously, this procedure to determine the Fourier coefficients is not computationally tractable. The size of the truth table is exponential in the number of variables (2^n), and thus the graph is going to be exponential in the number of variables. Fortunately, there are a variety of significantly faster methods for computing the Fourier basis of boolean functions [118]. One approach is to use the *Hadamard* matrix, which is defined on n variables as

$$H_n = H_{n-1} \otimes \begin{bmatrix} 1 & 1 \\ 1 & -1 \end{bmatrix},$$

where H_1 is defined by the matrix on the right. Fast Hadamard transforms have been developed that can take a compact representation of a boolean function specified using an *algebraic decision diagram* (ADD), and compute the Fourier coefficients in time that is polynomial in the size of the decision diagram.

Fourier representations have played a significant role in work on learning boolean functions. It was first shown that boolean functions can be efficiently learned under the uniform distribution by estimating the Fourier coefficients of the low-order basis functions [71]. In contrast, another Fourier learning algorithm was developed that could find the largest Fourier coefficients $> \theta$ in time polynomial in $\frac{n}{\theta}$ for a boolean function over n variables [65, 84]. This second algorithm is recursive, and requires a membership oracle.

Let $f : \{0, 1\}^n \to \{-1, 1\}$ be a boolean function over n variables. The Fourier transform of f is defined as

$$\hat{f}(\alpha) = \frac{1}{2^n} \sum_{x \in \{0,1\}^n} f(x)\chi_\alpha(x),$$

where χ_α is the Fourier basis function that computes the *parity* of a given function f on input x with respect to the string α. In other words, there are 2^n basis functions over all n-length boolean strings α, such that $\chi_\alpha(x) = +1$ if $\sum_{i=0}^n \alpha(i)x(i)$ is even, otherwise $\chi_\alpha(x) = -1$. The Fourier basis functions χ_α can be viewed as the group characters of the direct product of n Abelian two-element groups $\{+1, -1\}$ (note that the character of a direct product of two groups is the product of the characters of each group). Also, the expression above can be written more compactly simply as

$$\hat{f}(\alpha) = \langle f, \chi_\alpha \rangle,$$

where the inner product is with respect to the product group. The original function f can be reconstituted from its Fourier coefficients in the analysis–synthesis tradition as

$$f(x) = \sum_{\alpha \in \{0,1\}^n} \hat{f}(\alpha)\chi_\alpha(x).$$

There is an interesting connection between work in compressed sensing and Fourier representations of boolean functions. In both cases, the aim is to discover a sparse representation of the original function f by finding the (largest) coefficients of the Fourier representation by a randomized method. Representation discovery in this setting is not the computation of a basis, but rather, the computation of a sparse representation with respect to a given (but perhaps exponentially large) basis. More practical variants of Fourier boolean methods are described in [39], with promising experimental results on some data sets from the UCI Repository. The use of Fourier and wavelet bases with rich representations, such as first-order logic, is a significantly underdeveloped area, with much work remaining to be done.

9.3   GROUP REPRESENTATION THEORY

Harmonic analysis is often called the study of *symmetry* [50]. For reasons of space, we could not provide a detailed discussion of the intimate relationship between harmonic analysis and the representation theory of groups [51, 105]. Group representation theory builds on matrix representation theory in many ways. The *representation* of a group studies the action of a group on a vector space. It is possible to define Fourier analysis abstractly on groups [115], and the abstract Fourier expansion on the Hilbert space of functions on a group is similar to that used in this book. Classical Fourier analysis emerges as a special case where the group in question is the unit circle (or in the discrete setting, a ring graph).

Groups are often categorized into two types: in *Abelian* groups, the group operator is commutative (an example is the group of integers modulo n, under addition). Classical Fourier analysis as well as spectral analysis of time-series is largely in the setting of Abelian groups [38]. However, in many applications, such as the group of automorphisms of a graph, motion planning in robotics or search problems such as Rubik's cube, the group operator is not commutative (a translation in 3D space followed by a rotation does not yield the same outcome as doing these operations in the opposite order).

Commutative harmonic analysis is restricted to Abelian groups, whereas *non-commutative harmonic analysis* extends the setting to non-Abelian groups [25]. In commutative harmonic analysis, functions on a group can be decomposed into *irreducible* one-dimensional representations. These are essentially like eigenvalues (and their associated eigenvectors). In the more general setting of non-commutative harmonic analysis, the irreducible representations are no longer one-dimensional and form subspaces.

Group theory provides a powerful tool for combating large spaces. One approach is to decompose the space of functions on symmetric graphs with large automorphism groups using *character tables*. A group character is the trace of the matrix representation of a group. The character table lists the irreducible representations of a group associated with the conjugate classes of the group. Another property we can exploit is that random walk and Laplacian operators commute under graph automorphisms.

As a concrete illustration of the group-theoretic perspective, we show how the space of functions on a graph can be decomposed by exploiting symmetries of the graph [31, 61]. Given an undirected graph $G = (V, E, W)$, an *automorphism* π of a graph is a bijection $\pi : V \to V$ that leaves the weight matrix invariant. In other words, $w(u, v) = w(\pi(u), \pi(v))$. An automorphism π can be also represented in matrix form by a permutation matrix Γ that commutes with the weight matrix:

$$\Gamma W = W\Gamma.$$

The set of all such automorphisms forms a group, which is obviously a subgroup of the symmetric group $S_{|V|}$. Consider the graph shown in Figure 3.1. The adjacency (or Laplacian) matrix of this graph is invariant to rotations by multiples of 45 degrees, or reflections on the horizontal, vertical, or diagonal axes. The set of all such transformations represents the *dihedral*

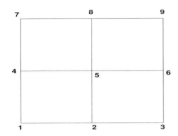 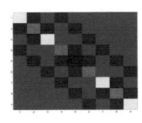 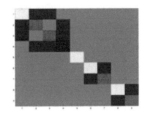

FIGURE 9.3: Decomposition of the combinatorial Laplacian operator on the grid graph (left) by exploiting the induced group of automorphisms (rotations and reflections). The original Laplacian matrix is shown in the middle and the reduced matrix on the right.

group [105]. These automorphisms leave the valency, or degree of a vertex, invariant, and consequently, the Laplacian is invariant under an automorphism. The set of all automorphisms forms a non-Abelian group, in that the group operation is non-commutative. Let x be an eigenvector of the combinatorial graph Laplacian L. Then, it is easy to show that Γx must be an eigenvector as well for any automorphism Γ. This result follows because

$$L\Gamma x = \Gamma L x = \Gamma \lambda x = \lambda \Gamma x.$$

It can be shown that the automorphisms of a graph factorize the graph Laplacian into irreducible blocks [61]. There is a close relationship between the abstract group induced by the automorphism group of a graph, and the spectral structure of the Laplacian matrix. One example of a key result is stated below [31].

Theorem 9.1.   *Given a graph G whose Laplacian $L(G) = D - A$ is such that no eigenvalue is repeated, the abstract group induced by the set of all automorphisms that commute with the Laplacian is Abelian (because all automorphisms Γ are involutions, that is $\Gamma^2 = I$).*

The proof of this theorem follows readily from the observation that if an eigenvalue is simple (of geometric multiplicity $= 1$), then the corresponding eigenvector x and Γx must be dependent.

It can be shown that the permuted eigenvector Γx is independent of the original eigenvector x if the corresponding eigenvalue λ is of geometric multiplicity > 1 [31]. It is possible to exploit the theory of linear representations of groups to decompose operators on graphs, such as shown in Figure 9.3. The use of the Laplacian in constructing representations that are *invariant* to group operations is a hallmark of work in harmonic analysis [50].

9.4 BIBLIOGRAPHICAL REMARKS

Compressed sensing is a very active area of research. The web page `http://www.dsp.ece.rice.edu/cs/` contains a detailed overview of this rapidly growing field, including software packages. Figure 9.1 was based on the MATLAB package l1Magic. Bernasconi's PhD dissertation [13] has a detailed discussion of the connection between the graph Laplacian and the Fourier representation of boolean functions. For a detailed introduction to group representation theory, see the classic texts by Hamermesh [51] and Serre [105].

Bibliography

[1] D. Achlioptas, F. McSherry, and B. Scholkopff, "Sampling techniques for kernel methods," in *Proceedings of the 14th International Conference on Neural Information Processing Systems (NIPS)*, Cambridge, MA: MIT Press, pp. 335–342, 2002.

[2] S. Amarel, "On representations of problems of reasoning about actions," in *Machine Intelligence*, Vol. 3, D. Michie, Ed. Amsterdam: Elsevier/North-Holland, 1968, pp. 131–171.

[3] N. Amhed, T. Natarajan, and K. Rao, "On image processing and a discrete cosine transform," *IEEE Transactions on Computers*, C-23(1):90–93, 1974. doi:10.1109/T-C.1974.223784

[4] S. Axler, *Linear Algebra Done Right*. Berlin: Springer, 1997.

[5] S. Axler, P. Bourdon, and W. Ramey, *Harmonic Function Theory*. Berlin: Springer, 2001.

[6] J. Bagnell and J. Schneider, "Covariant policy search," in *Proceedings of the International Joint Conference on Artificial Intelligence (IJCAI)*, pp. 1019–1024, 2003.

[7] C. Baker, *The Numerical Treatment of Integral Equations*. Oxford: Clarendon Press, 1977.

[8] A. Barto and S. Mahadevan, "Recent advances in Hierarchical Reinforcement Learning," *Discrete Event Systems Journal*, 13:41–77, 2003. doi:10.1023/A:1022140919877

[9] M. Belkin and P. Niyogi, "Laplacian Eigenmaps for dimensionality reduction and data representation," *Neural Computation*, 6(15):1373–1396, June 2003. doi:10.1162/089976603321780317

[10] M. Belkin and P. Niyogi, "Semi-supervised learning on Riemannian manifolds," *Machine Learning*, 56:209–239, 2004. doi:10.1023/B:MACH.0000033120.25363.1e

[11] M. Belkin and P. Niyogi, "Towards a theoretical foundation for Laplacian-based manifold methods," in *Proceedings of the International Conference on Computational Learning Theory (COLT)*, pp. 486–500, 2005.

[12] S. Belongie, C. Fowlkes, F. Chung, and J. Malik, "Spectral partitioning with indefinite kernels using the Nyström extension," in *Proceedings of the 7th European Conference on Computer Vision*, pp. 531–542, 2002.

[13] A. Bernasconi, *Mathematical Techniques for Analysis of Boolean Functions*. PhD thesis, University of Pisa, 1998.

[14] D. Bertsekas and J. Tsitsiklis, *Neuro-Dynamic Programming*. Belmont, MA: Athena Scientific, 1996.

[15] L. Billera and P. Diaconis, "A geometric interpretation of the Metropolis–Hasting algorithm," *Statistical Science*, 16:335–339, 2001. doi:10.1214/ss/1015346318

[16] C. Bishop, *Machine Learning and Pattern Recognition*. Berlin: Springer, 2006.

[17] D. Blei, T. Griffiths, M. Jordan, and J. Tenenbaum, "Hierarchical topic models and the nested Chinese restaurant process," in *Proceedings of the International Conference on Neural Information Processing Systems*, 2004.

[18] D. Blei, A. Ng, and M. Jordan, "Latent Dirichlet Allocation," *Journal of Machine Learning Research*, 3:993–1022, 2003.

[19] S. Bradtke and A. Barto, "Linear least-squares algorithms for temporal difference learning," *Machine Learning*, 22:33–57, 1996.

[20] A. Brandt, "Algebraic multigrid theory: The symmetric case," *Applied Mathematics and Computation*, 19(1–4):23–56, 1986.

[21] J. Bremer, R. Coifman, M. Maggioni, and A. Szlam, "Diffusion wavelet packets," *Applied and Computational Harmonic Analysis*, 21(1):95–112, July 2006. doi:10.1016/j.acha.2006.04.005

[22] E. Candes, J. Romberg, and T. Tao, "Robust uncertainty principles: Exact signal reconstruction from highly incomplete frequency information," *IEEE Transactions on Information Theory*, 52(2):489–509, 2006. doi:10.1109/TIT.2005.862083

[23] I. Chavel, *Eigenvalues in Riemannian Geometry*. Pure and Applied Mathematics. New York: Academic, 1984.

[24] J. Cheeger, "A lower bound for the smallest eigenvalue of the Laplacian," in *Problems in Analysis*, R. C. Gunning, Ed., pp. 195–199. Princeton: Princeton University Press, 1970.

[25] G. Chirikjian and A. Kyatkin, *Engineering Applications of Noncommutative Harmonic Analysis*. Boca Raton, FL: CRC Press, 2001.

[26] F. Chung, *Spectral Graph Theory*. Number 92 in CBMS Regional Conference Series in Mathematics. American Mathematical Society, 1997. doi:10.1007/s00026-005-0237-z

[27] F Chung, Laplacians and the Cheeger Inequality for Directed Graphs. *Annals of Combinatorics*, 9(1):1–19, April 2005.

[28] R. Coifman, S. Lafon, A. Lee, M. Maggioni, B. Nadler, F. Warner, and S. Zucker, "Geometric diffusions as a tool for harmonic analysis and structure definition of data: part I. Diffusion maps," *Proceedings of National Academy of Science*, 102(21):7426–7431, May 2005. doi:10.1073/pnas.0500334102

[29] R. Coifman, S. Lafon, A. Lee, M. Maggioni, B. Nadler, F. Warner, and S. Zucker, "Geometric diffusions as a tool for harmonic analysis and structure definition of data: part II.

Multiscale methods," *Proceedings of the National Academy of Science*, 102(21):7432–7437, May 2005. doi:10.1073/pnas.0500896102

[30] R. Coifman and M. Maggioni, "Diffusion wavelets," *Applied and Computational Harmonic Analysis*, 21(1):53–94, July 2006. doi:10.1016/j.acha.2006.04.004

[31] D. Cvetkovic, M. Doob, and H. Sachs, *Spectra of Graphs: Theory and Application*. New York: Academic, 1980.

[32] D. Cvetkovic, P. Rowlinson, and S. Simic, *Eigenspaces of Graphs*. Cambridge: Cambridge University Press, 1997.

[33] T. Das, A. Gosavi, S. Mahadevan, and N. Marchalleck, "Solving semi-markov decision problems using average-reward reinforcement learning," *Management Science*, 45(4):560–574, 1999.

[34] I. Daubechies, *Ten Lectures on Wavelets*. Society for Industrial and Applied Mathematics, 1992.

[35] P. Dayan, "Improving generalisation for temporal difference learning: The successor representation," *Neural Computation*, 5:613–624, 1993. doi:10.1162/neco.1993.5.4.613

[36] S. Deerwester, S. Dumais, G. Furnas, T. Landauer, and R. Harshman, "Indexing by latent semantic analysis," *Journal of the American Society for Information*, 1990.

[37] F. Deutsch, *Best Approximation in Inner Product Spaces*. Canadian Mathematical Society, 2001.

[38] P. Diaconis, *Group Representations in Probability and Statistics*. Institute of Mathematical Statistics, 1988.

[39] A. Drake and D. Ventura, "A practical generalization of Fourier-based learning," in *ICML '05: Proceedings of the 22nd International Conference on Machine Learning*, pp. 185–192, New York, NY, USA, 2005. ACM.

[40] P. Drineas, R. Kannan, and M. W. Mahoney, "Fast Monte Carlo algorithms for matrices II: Computing a low-rank approximation to a matrix," Technical Report YALEU/DCS/TR-1270, Yale University Department of Computer Science, New Haven, CT, February 2004.

[41] P. Drineas and M. W. Mahoney, "On the Nyström method for approximating a Gram matrix for improved kernel-based learning," *Journal of Machine Learning and Research*, 6:2153–2175, 2005.

[42] C. Drummond, "Accelerating reinforcement learning by composing solutions of automatically identified subtasks," *Journal of AI Research*, 16:59–104, 2002.

[43] Ebay, EBay IT Discussion Board, http://forums.ebay.com/.

[44] M. Fiedler, "Algebraic connectivity of graphs," *Czech. Math. Journal*, 23(98):298–305, 1973.

[45] D. Foster and P. Dayan, "Structure in the space of value functions," *Machine Learning*, 49:325–346, 2002. doi:10.1023/A:1017944732463

[46] A. Frieze, R. Kannan, and S. Vempala, "Fast Monte Carlo algorithms for finding low-rank approximations," in *Proceedings of the 39th Annual IEEE Symposium on Foundations of Computer Science*, pp. 370–378, 1998.

[47] G. Golub and C. Van Loan, *Matrix Computations*. Baltimore, MD: John Hopkins University Press, 1989.

[48] W. Grassmann, M. Taksar, and D. Heyman, "Regenerative analysis and steady state distributions for Markov Chains," *Operations Research*, 33(5):1107–1116, 1985.

[49] C. Guestrin, A. Krause, and A. Singh, "Near-optimal sensor placements in Gaussian processes," in *22nd International Conference on Machine Learning*, July 2005.

[50] D. Gurarie, *Symmetries and Laplacians: Introduction to Harmonic Analysis, Group Representations and Laplacians*. Amsterdam: North-Holland, 1992.

[51] M. Hamermesh, *Group Theory and its Application to Physical Problems*. New York: Dover, 1989.

[52] M. Hein, J. Audibert, and U. von Luxburg, "Graph Laplacians and their convergence on random neighborhood graphs," *Journal of Machine Learning Research*, 8:1325–1368, 2007.

[53] T. Hofmann, "Probabilistic latent semantic indexing," in *Proceedings of the 22nd Annual International SIGIR Conference*, 1999.

[54] J. Jackson, *The Harmonic Sieve: A Novel Application of Fourier Analysis to Machine Learning Theory and Practice*. PhD thesis, Carnegie-Mellon University, 1995.

[55] J. Johns and S. Mahadevan, "Constructing basis functions from directed graphs for value function approximation," in *Proceedings of the International Conference on Machine Learning (ICML)*, pp. 385–392. ACM Press, 2007.

[56] J. Johns, S. Mahadevan, and C. Wang, "Compact spectral bases for value function approximation using Kronecker Factorization," in *Proceedings of the National Conference on Artificial Intelligence (AAAI)*, 2007.

[57] T. Jolliffe, *Principal Components Analysis*. Berlin: Springer, 1986.

[58] S. Kakade, "A natural policy gradient," In *Proceedings of Neural Information Processing Systems*. Cambridge, MA: MIT Press, 2002.

[59] Z. Karni and C. Gotsman, "Spectral compression of mesh geometry," in *SIGGRAPH '00: Proceedings of the 27th Annual Conference on Computer Graphics and Interactive Techniques*, pp. 279–286. ACM Press/Addison-Wesley Publishing Co., 2000. doi:10.1145/344779.344924

[60] G. Karypis and V. Kumar, "A fast and high quality multilevel scheme for partitioning irregular graphs," *SIAM Journal of Scientific Computing*, 20(1):359–392, 1999.

[61] A. Kaveh and A. Nikbakht. Block diagonalization of Laplacian matrices of symmetric graphs using group theory. *International Journal for Numerical Methods in Engineering*, 69:908–947, 2007. doi:10.1002/nme.1794

[62] P. Keller, S. Mannor, and D Precup, "Automatic basis function construction for approximate dynamic programming and reinforcement learning," in *Proceedings of the 22nd International Conference on Machine Learning (ICML)*, pp. 449–456. Cambridge, MA: MIT Press, 2006.

[63] D. Koller and R. Parr, "Policy iteration for factored MDPs," in *Proceedings of the 16th Conference on Uncertainty in AI*, pp. 326–334, 2000.

[64] I. Koutis and G. Miller, "A linear work, $O(n^{1/6})$ time, parallel algorithm for solving planar Laplacians," in *Symposium on Discrete Algorithms (SODA)*, pp. 1002–1011, 2007.

[65] E. Kushilevitz and Y. Mansour, "Learning decision trees using the Fourier spectrum," In *STOC '91: Proceedings of the 23rd Annual ACM Symposium on Theory of Computing*, pp. 455–464, New York, NY, USA, 1991. ACM. doi:10.1145/103418.103466

[66] J. Lafferty and G. Lebanon, "Diffusion kernels on statistical manifolds," *Journal of Machine Learning Research*, 6:129–163, 2005.

[67] M. Lagoudakis and R. Parr, "Least-squares policy iteration," *Journal of Machine Learning Research*, 4:1107–1149, 2003. doi:10.1162/jmlr.2003.4.6.1107

[68] J. C. Latombe, *Robot Motion Planning*. Dordrecht: Kluwer, 1991.

[69] S. Lavalle, *Planning Algorithms*. Cambridge: Cambridge University Press, 2006.

[70] J. M. Lee, *Introduction to Smooth Manifolds*. Berlin: Springer, 2003.

[71] N. Linial, Y. Mansour, and N. Nisan, "Constant depth circuits, Fourier transform, and learnability," *Journal of the ACM*, 40(3):607–620, 1993. doi:10.1145/174130.174138

[72] M. Maggioni and S. Mahadevan, "Fast direct policy evaluation using multiscale analysis of Markov diffusion processes," in *Proceedings of the 23rd International Conference on Machine Learning*, pp. 601–608, New York, NY, USA, 2006. ACM Press. doi:10.1016/0004-3702(92)90058-6

[73] M. Maggioni and S. Mahadevan, A multiscale framework for markov decision processes using diffusion wavelets, Technical Report TR-2006-36, Department of Computer science, University of Massachusetts, 2006.

[74] S. Mahadevan, "Proto-value functions: developmental reinforcement learning," in *Proceedings of the International Conference on Machine Learning*, pp. 553–560, 2005.

[75] S. Mahadevan, "Representation policy iteration," in *Proceedings of the 21th Annual Conference on Uncertainty in Artificial Intelligence (UAI-05)*, pp. 372–37. AUAI Press, 2005.

[76] S. Mahadevan, "Adaptive mesh compression in 3D computer graphics using multiscale manifold learning," in *Proceedings of the International Conference on Machine Learning (ICML)*, pp. 585–592. ACM Press, 2007.

[77] S. Mahadevan and J. Connell, "Automatic programming of behavior-based robots using reinforcement learning," *Artificial Intelligence*, 55:311–365, 1992 (appeared originally as IBM TR RC16359, Dec 1990).

[78] S. Mahadevan and M. Maggioni, "Value function approximation with Diffusion Wavelets and Laplacian Eigenfunctions," in *Proceedings of the Neural Information Processing Systems (NIPS)*. Cambridge, MA: MIT Press, 2006.

[79] S. Mahadevan and M. Maggioni. Proto-Value Functions: A Laplacian Framework for Learning Representation and Control in Markov Decision Processes. *Journal of Machine Learning Research*, 8:2169–2231, 2007.

[80] S. Mahadevan, M. Maggioni, K. Ferguson, and S. Osentoski, "Learning representation and control in continuous markov decision processes," in *Proceedings of the National Conference on Artificial Intelligence (AAAI)*, 2006.

[81] S. Mallat, "A theory for multiresolution signal decomposition: The wavelet representation," *IEEE Transactions on Pattern Analysis and Machine Intelligence*, 11(7):674–693, 1989. doi:10.1109/34.192463

[82] S. Mallat, *A Wavelet Tour in Signal Processing*. New York: Academic, 1998.

[83] C. Manning and H. Schütze, *Foundations of Statistical Natural Language Processing*. Cambridge, MA: MIT Press, 1999.

[84] Y. Mansour and S. Sahar, Implementation issues in the Fourier transform algorithm, *Machine Learning*, 40(1):5–33, 2000. doi:10.1023/A:1011034100370

[85] A. McCallum, A. Corrada-Emmanuel, and X. Wang, The author-recipient-topic model for topic and role discovery in social networks: Experiments with enron and academic email, Technical Report UM-CS-2004-096, Department of Computer Science, University of Massachusetts, Amherst, 2004.

[86] M. Meila and J. Shi, "Learning segmentation by random walks," in *NIPS*, 2001.

[87] C. Meyer, "Uncoupling the Perron Eigenvector Problem," *Linear Algebra and its Applications*, 114/115:69–94, 1989. doi:10.1016/0024-3795(89)90452-7

[88] A. Ng, M. Jordan, and Y. Weiss, "On spectral clustering: Analysis and an algorithm," in *Proceedings of the Neural Information Processing Systems*, 2002.

[89] A. Ng, H. Kim, M. Jordan, and S. Sastry, "Autonomous helicopter flight via reinforcement learning," in *Proceedings of Neural Information Processing Systems*, 2004.

[90] P. Niyogi and M. Belkin, Semi-supervised learning on Riemannian manifolds, Technical Report TR-2001-30, University of Chicago, Computer Science Dept., Nov. 2001.

[91] P. Niyogi, I. Matveeva, and M. Belkin, Regression and regularization on large graphs. Technical report, University of Chicago, Nov. 2003.

[92] S. Osentoski and S. Mahadevan, "Learning State Action Basis Functions for Hierarchical Markov Decision Processes," in *Proceedings of the International Conference on Machine Learning (ICML)*, pp. 705–712, 2007.

[93] R. Parr, C. Painter-Wakefiled, L. Li, and M. Littman, "Analyzing feature generation for value function approximation," in *Proceedings of the International Conference on Machine Learning (ICML)*, pp. 737–744, 2007.

[94] J. Peters, S. Vijaykumar, and S. Schaal, "Reinforcement learning for humanoid robots," in *Proceedings of the 3rd IEEE-RAS International Conference on Humanoid Robots*, 2003.

[95] M. Petrik, "An analysis of Laplacian methods for value function approximation in MDPs," in *Proceedings of the International Joint Conference on Artificial Intelligence (IJCAI)*, pp. 2574–2579, 2007.

[96] J. C. Platt, FastMap, MetricMap, and Landmark MDS are all Nyström algorithms, Technical Report MSR-TR-2004-26, Microsoft Research, Sep. 2004.

[97] P. Poupart and C. Boutilier, "Value directed compression of POMDPs," in *Proceedings of the International Conference on Neural Information Processing Systems (NIPS)*, 2003.

[98] M. L. Puterman, *Markov Decision Processes*. Wiley Interscience, New York, USA, 1994.

[99] S. Rosenberg, *The Laplacian on a Riemannian Manifold*. Cambridge: Cambridge University Press, 1997.

[100] S. Roweis. Neural information processing systems NIPS 1–12 papers data set. http://www.cs.toronto.edu/~roweis/data.html. doi:10.1126/science.290.5500.2323

[101] S. Roweis and L. Saul, Nonlinear dimensionality reduction by local linear embedding. *Science*, 290:2323–2326, 2000.

[102] S. Russell and P. Norvig, *Artificial Intelligence: A Modern Approach*. Englewood Cliffs, NJ: Prentice-Hall, 2002.

[103] B. Sallans and G. Hinton, "Reinforcement learning with factored states and actions," *Journal of Machine Learning Research*, 5:1063–1088, 2004.

[104] B. Scholkopf and A. Smola, *Learning with Kernels: Support Vector Machines, Regularization, Optimization, and Beyond*. Cambridge, MA: MIT Press, 2001.

[105] J. Serre, *Linear Representations of Finite Groups*. Berlin: Springer, 1977.

[106] J. Shi and J. Malik, "Normalized cuts and image segmentation," *IEEE PAMI*, 22:888–905, 2000.

[107] O. Sorkine, D. Cohen-Or, D. Irony, and S. Toledo, Geometry-aware bases for shape approximation, *IEEE Transactions on Visualization and Computer Graphics*, 11(2):171–180, 2005. doi:10.1109/TVCG.2005.33

[108] D. Spielman and S. Teng, "Nearly-linear time algorithms for graph partition-
       ing, graph sparsification, and solving linear systems," in *STOC '04: Proceedings
       of the 36th Annual ACM Symposium on Theory of Computing*, pp. 81–90, 2004.
       doi:10.1145/1007352.1007372

[109] G. Stewart and J. Sun, *Matrix Perturbation Theory*. New York: Academic, 1990.

[110] G. Strang, *Introduction to Linear Algebra*. Cambridge, MA: Wellesley-Cambridge Press,
       2003.

[111] R. Sutton and A. G. Barto, *An Introduction to Reinforcement Learning*. Cambridge, MA:
       MIT Press, 1998.

[112] G. Taubin, "A signal processing approach to fair surface design," in *SIGGRAPH '95:
       Proceedings of the 22nd Annual Conference on Computer Graphics and Interactive Techniques*,
       pp. 351–358. ACM Press, 1995. doi:10.1145/218380.218473

[113] G. Taubin, T Zhang, and G. Golub, Optimal surface smoothing as filter design, in
       ECCV (1), pp. 283–292, 1996.

[114] J. Tenenbaum, V. de Silva, and J. Langford, A global geometric frame-
       work for nonlinear dimensionality reduction, *Science*, 290:2319–2323, 2000.
       doi:10.1126/science.290.5500.2319

[115] A. Terras, *Fourier Analysis on Finite Groups and Applications*. Cambridge: Cambridge
       University Press, 1999.

[116] G. Tesauro, "TD-Gammon, a self-teaching backgammon program, achieves master-
       level play," *Neural Computation*, 6:215–219, 1994. doi:10.1162/neco.1994.6.2.215

[117] G. Theocharous, K. Rohanimanesh, and S. Mahadevan, "Learning hierarchical partially
       observable markov decision processes for robot navigation," in *IEEE Conference on
       Robotics and Automation (ICRA)*, 2001.

[118] M. Thornton, R. Drechsler, and D. Miller, *Spectral Methods for VLSI Design*. Dordrecht:
       Kluwer, 2001.

[119] B. Turker, J. Leydold, and P. Stadler, *Laplacian Eigenvectors of Graphs*. Berlin: Springer,
       2007.

[120] C. Van Loan and N. Pitsianis, "Approximation with Kronecker Products," in *Linear
       Algebra for Large Scale and Real Time Applications*, pp. 293–314. Dordrecht: Kluwer,
       1993.

[121] B. Van Roy, *Learning and Value Function Approximation in Complex Decision Processes*.
       PhD thesis, MIT, 1998.

[122] G. Wallace, "The JPEG still picture compression standard," *Communications of the
       ACM*, 34(4):30–44, 1991. doi:10.1145/103085.103089

[123] C. Wang and S. Mahadevan, "Multiscale Analysis of Document Corpora Based on
       Diffusion Models," University of Massachusetts, Amherst, Technical Report TR-2008-
       16, 2008.

[124] R. Wang, J. Traan, and D. Luebke, "All-frequency relighting of glossy objects," *ACM Transactions on Graphics*, 25(2), 2006.

[125] C. Watkins, *Learning from Delayed Rewards*. PhD thesis, King's College, Cambridge, England, 1989.

[126] C. Williams and M. Seeger, "Using the Nyström Method to speed up Kernel Machines," in *Proceedings of the International Conference on Neural Information Processing Systems*, pp. 682–688, 2000.

[127] W. Zhang and T. Dietterich, "A reinforcement learning approach to job-shop scheduling," in *Proceedings of the 14th International Joint Conference on Artificial Intelligence (IJCAI)*, pp. 1114–1120, 1995.

[128] X. Zhou, *Semi-Supervised Learning With Graphs*. PhD thesis, Carnegie Mellon University, 2005.

Author Biography

Dr. Sridhar Mahadevan is an Associate Professor in the Department of Computer Science at the University of Massachusetts, Amherst. He received his PhD from Rutgers University in 1990. Professor Mahadevan's research interests span several subfields of artificial intelligence and computer science, including machine learning, multi-agent systems, planning, perception, and robotics. His PhD thesis introduced the learning apprentice model of knowledge acquisition from experts, as well as a rigorous study of concept learning with prior determination knowledge using the framework of Probably Approximately Correct (PAC) learning. In 1993, he co-edited (with Jonathan Connell) the book Robot Learning published by Kluwer Academic Press, one of the first books on the application of machine learning to robotics. Over the past decade, his research has centered around Markov decision processes and reinforcement learning, where his papers are among the most cited in the field. His recent work on spectral and wavelet methods for Markov decision processes has generated much attention, leading to a unified framework for learning representation and behavior.

Professor Mahadevan is an Associate Editor for the *Journal of Machine Learning Research*. Previously, he served for many years as an Associate Editor for *Journal of AI Research* and the *Machine Learning Journal*. He has been on numerous program committees for AAAI, ICML, IJCAI, NIPS, ICRA, and IROS conferences, including area chair for at AAAI, ICML, and NIPS conferences. In 2001, he co-authored a paper with his students Rajbala Makar and Mohammad Ghavamzadeh that received the best student paper award in the 5th International Conference on Autonomous Agents. In 1999, he co-authored a paper with Gang Wang that received the best paper award (runner-up) at the 16th International Conference on Machine Learning. He was an invited tutorial speaker at ICML 2006, IJCAI 2007, and AAAI 2007.

Printed in the United States
by Baker & Taylor Publisher Services